Synthetic

Synthetic

How Life Got Made

SOPHIA ROOSTH

THE UNIVERSITY OF CHICAGO PRESS   CHICAGO AND LONDON

The University of Chicago Press, Chicago 60637
The University of Chicago Press, Ltd., London
© 2017 by The University of Chicago
All rights reserved. Published 2017.
Printed in the United States of America

26 25 24 23 22 21 20 19 18 17   1 2 3 4 5

ISBN-13: 978-0-226-44032-3 (cloth)
ISBN-13: 978-0-226-44046-0 (paper)
ISBN-13: 978-0-226-44063-7 (e-book)
DOI: 10.7208/chicago/9780226440637.001.0001

Library of Congress Cataloging-in-Publication Data

Names: Roosth, Sophia, author.
Title: Synthetic : how life got made / Sophia Roosth.
Description: Chicago ; London : The University of Chicago Press, 2017. |
    Includes bibliographical references and index.
Identifiers: LCCN 2016027488 | ISBN 9780226440323 (cloth : alk. paper) |
    ISBN 9780226440460 (pbk. : alk. paper) | ISBN 9780226440637 (e-book)
Subjects: LCSH: Synthetic biology. | Bioengineering.
Classification: LCC TA164 .R66 2017 | DDC 660/.6509—dc23 LC record available at
    https://lccn.loc.gov/2016027488

Contents

Analysis: Synthesis

L ife is not what it used to be. This is a story about how it got that way. Living things bearing genomes pared down, streamlined, or cobbled together from bits of synthesized DNA now scurry, swim, and flourish in test tubes and glass bioreactors: viruses named for computer software, bacteria encoding passages of James Joyce, chimeric yeast buckling under the metabolic strain of genes harvested from sweet wormwood, petunias, and microbes from Icelandic thermal pools.

In the final years of the twentieth century, émigrés from mechanical and electrical engineering and computer science resolved that if the aim of biology was to understand life, then *making* life would yield better theories than would experimentation. Many of these researchers clustered at the Massachusetts Institute of Technology (MIT) beginning in 2003. Naming themselves synthetic biologists, they advocate not experiment but manufacture, not reduction but construction, not analysis but synthesis. As a cultural anthropologist, I spent eight years talking to and working with them.[1]

Armed with biotechnology techniques—notably, faster and cheaper methods for DNA sequencing and synthesis—this new breed of life scientists treats biological media as a substrate for manufacture, raw material that can be manipulated using engineering principles borrowed from their various home disciplines. Sequencing and synthesis allow synthetic biologists to traffic between physical molecules of nucleic acid (DNA and RNA) and dematerialized genetic sequences scrolling across computer screens. Sequencing means "reading" the strings of four nucleotide bases whose sequence constitutes DNA and RNA to compose a digital genetic "code" made up entirely of letters that stand in for the molecule (A for the nucleotide adenine, C for cytosine, G for guanine, T for thymine).[2] Synthesis does the reverse: using elaborate genomic techniques, researchers can physically build material nucleic acid macromolecules to order on the basis of desired genetic codes. Beyond these tools, work in a synthetic biology lab is almost indistinguishable from that in most molecular biology labs. Synthetic biology's distinctiveness is more apparent in its approach and its practitioners' speech than in the day-to-day benchwork of its labs.[3]

Two synthetic biologists define their field: "Synthetic biologists seek to assemble components that are not natural (therefore synthetic) to generate chemical systems that support Darwinian evolution (therefore biological). By carrying out the assembly in a synthetic way, these scientists hope to understand non-synthetic biology, that is, 'natural' biology."[4] In their equation of making with understanding, of synthesis with analysis, making life is not an end in itself but rather a technique for probing life's margins. Making new life-forms also requires that researchers query seemingly commonsense terms like "natural" and "unnatural," "biological" and "synthetic."

By 2010, some seven years after its origins at MIT, the synthetic biology research economy had swelled considerably. As of 2006, its overall annual market was US$600 million, and between 2005 and 2010 alone, synthetic biology researchers received over $430 billion from the US government.[5] The vast majority of synthetic biology research today occurs in Western Europe (especially the United Kingdom, Germany, and the Netherlands) and the United States (where most work is concentrated in California and New England, and Massachusetts in particular). Still, there is a thin spread of synthetic biology research across the country: in 2013 an estimated 174 American universities reported that they engaged in some kind of synthetic biology research.[6] This research is primarily funded federally by the National Institutes of Health, the Defense Advanced Research Projects Agency (DARPA), the Department of Energy, and the National Science Foundation, as well as by private organizations, universities, and venture capital.[7]

Such organizations, which fund the lion's share of professional synthetic biology research, are interested in the field primarily for its potential commercial applications: clean energy, bioweapons, and cheap drug synthesis.[8] Such work often bears practical resemblances to biotechnology, synthetic chemistry, chemical engineering, and pharmaceuticals research. The most prominent—and mediagenic—example of synthetic biology is Professor Jay Keasling's (University of California, Berkeley) development of a synthetic microbial pathway to manufacture inexpensive artemisinin, an anti-malarial compound, for distribution in developing countries.[9]

However, the difference between synthetic biologists' impulse and earlier examples of biological experimentation is that they do not make living things in the service of discovery science or experimental research alone. Rather, making is also an *end* in itself. Newly built biotic things serve as answers to biological questions that might otherwise have remained unasked. They are tools with which synthetic biologists theorize what life is.

One notable synthetic biology project, for example, was aimed more at developing new synthesis techniques than a product. In 2010 synthetic biologists at the J. Craig Venter Institute (JCVI) synthesized a "minimal" organism,[10] a single-celled, independently living entity that maintains, JCVI researchers posit, the least genetic material necessary to sustain life.[11] JCVI announced that it had synthesized and assembled a synthetic version of the bacterium *Mycoplasma mycoides* genome, inserting it into *Mycoplasma capricolum* cells. Craig Venter exulted that his scientific team had made "the first self-replicating species we've had on the planet whose parent is a computer."[12] If synthetic biologists now define life according to its material construction, then a minimal organism is an ontological receding horizon: it is not the simplest, smallest, or most genetically austere organism found on Earth but rather the most genetically minimal viable organism that synthetic biologists can build. Projects such as this one toy with what counts as the "limits of life,"[13] from reverse engineering what researchers imagine to be the "first" life-form to stripping down microbes to determine a "minimal" genome.

This belief that biology can be understood through its construction is unmistakable not only in synthetic biologists' engineering projects but also in how they talk about their goals. Rob Carlson, a synthetic biologist and garage biohacker, makes a comparison to engineering: "Understanding is defined by the ability to build something new that behaves as expected," whether a 777 jet or a yeast cell.[14] Synthetic biologist Andrew "Drew" Endy uses a similar analogy to physics to make the same point: "Imagine what the science around the origin of the universe might be like if physicists could construct universes. It just so happens that in biology, the technology of synthesis [allows you to] instantly take your hypothesis and compile it into a physical instance and then test it."[15]

While working among MIT synthetic biologists, I learned their origin story for synthetic biology.[16] In the mid-1990s Endy, then an engineering graduate student, sought to build a software simulation of a simple virus called bacteriophage T7.[17] His computational model did not accurately predict the behavior of this virus (namely, the way it infected bacteria).

As a researcher and later a professor, he tasked his graduate students in 2004 not with reprogramming the software model but with rebuilding the physical virus *to conform to the software model.* Such a counterintuitive approach to making living things testifies to the lack of a coherent and meaningful substrate for identifying the *object* of biological inquiry. Engineered organisms now function as proofs of concept, demonstrations of accumulated biological knowledge, or signs of what about biology remains to be discovered. The logic goes like this: the best way to prove a biological theory or learn more about how life works is to "build" a new organism "from scratch."

Such reasoning recalls Artificial Life founder Chris Langton's disciplinary alibi from a decade earlier: "It's very difficult to build general theories about what life would be like anywhere in the universe and whatever it was made out of, when all we have to study is the unique example of life that exists here on Earth. So, what we have to do—perhaps—is the next best thing, which is to create far simpler systems in our computers."[18] The Artificial Life of the 1990s and the synthetic biology of the early 2000s have much in common, not the least of which being researchers' explicit efforts to build new instantiations of something they call "life." Nonetheless, Artificial Life was premised on abstracting life by simulating it in computer software. Artificial Life researchers treated life as if it were a universal formal category transcending substance, material, or medium. Something different motivates synthetic biologists' work. Most notably, they do not posit that life is a property separable from biological matter. Neither is their project mimetic: rather than imitate life, they construct new living kinds. Life's synthesis, in synthetic biology, forwards biological analysis, even as analysis conditions what gets synthesized and why.

Many synthetic biologists quote Richard Feynman, who scribbled on his Caltech blackboard just before his death: "What I cannot create, I do not understand."[19] This quotation is "a favorite among synthetic biologists—and for good reason," because "synthetic biology is the pursuit of comprehending biological systems by trying to engineer them."[20] Feynman's phrase would later be misquoted when coded and inserted into the genome of one of Venter's synthesized microbes. Endy also paraphrases Feynman when he says, "As a biological designer, until I can actually design something, I don't understand it."[21] Making has operated hand in glove with knowing since seventeenth-century Baconian mechanical philosophy dispensed with natural philosophy to experiment on the natural world. Experimentalists and artisans joined *theoria* to *practica,* and contemplation served instrumentation. Yet the ubiquity of "maker's knowledge" and artisanship in modern science has since largely been forgotten, especially in the mid- to late twentieth century, when scientific disciplines were divided into "pure" and "applied" research.[22] Synthetic biology is the latest instantiation of a centuries-long debate as to whether nature may be known through artifice.

Can genomes be "refactored" and streamlined to function like software code? Yes, synthetic biologists answer, because we have generated just such a bacteriophage (chapter 1). Can a living thing be fragmented into parts, and from a library of parts, can an organism be assembled? Yes, they say, because we have made standardized biological parts (chapters 2 and 3). What is the minimal system that is viable and free-living? The one we ourselves have made, they respond (chapter 3). Can species be defined beyond the continuous unspooling of biological generations? Yes, because we can revive species already extinct (chapter 6). Synthetic biologists make new living things in order better to understand how life works. Yet making recursively loops theory: the new living things biologists make function as "persuasive objects" that materialize theories of what synthetic biologists seek to understand about life. In short, the biological features, theories, and limits that synthetic biologists fasten upon are circularly determined by their own experimental tactics, which they then identify with the things they have made.

How did this astonishing avowal that biological making fosters biological knowing arise? How did synthetic biologists come to suture making to knowing? And how is that stance indebted to a longer history of biology? Biology has always been, since its inception and by definition, an inquiry into what life is. Michel Foucault claimed that "life itself" is a category that "did not exist" prior to the end of the eighteenth century: "Life does not constitute an obvious threshold beyond which entirely new forms of knowledge are required. It is a category of classification, relative, like all the other categories, to the criteria one adopts."[23] That is, biology as a discipline was warranted by a classificatory decision: carving up the world into the organic and the inorganic, differentiating between the vital and the lifeless, and insisting that the living world demanded a science of its own.

Biology, the just-so story goes, emerged at the start of the nineteenth century as a professional discipline from allied projects such as anatomy, taxonomy, and *Naturphilosophie*.[24] The Romantics and later the Victorians inductively studied living form, asking how living things are shaped by their surrounding habitats.[25] Darwin proposed his theory of natural selection, stimulating studies of speciation that would lead to the evolutionary synthesis in the twentieth century.[26] French and German physiologists moved things indoors. As biologists professionalized at the end of the nineteenth century, the life sciences became primarily experimental, using laboratory and field research to inquire into the structure and function of life. Biology's newfound locale was the laboratory, which rendered the experimental method both controllable and manageable.[27] Experimental apparatuses flattened lively processes into quantitative information and graphic traces.[28] Living things were literally turned inside out, as experimental biologists mimicked or artificially re-created their fluid internal milieux in glassware.[29]

Perhaps the closest philosophical forebear of synthetic biology was the German American biologist Jacques Loeb, who "considered the main problem of biology to be the production of the new, not the analysis of the existent."[30] In 1899 he induced sea urchin eggs to divide without fertilization, a technique termed "artificial parthenogenesis." News of his breakthrough made headlines, and single women wrote letters to Loeb in hopes that his new technique could separate human reproduction from heterosex. Loeb had little patience for "metaphysical" questions about life, which he considered a distraction from the ultimate aim of biology: control of the living world. This was not the first time new organisms had been made or manipulated (prominent earlier examples stretch from animal husbandry and the Neolithic origins of agriculture to the pigeon fanciers of Darwin's day). However, in the first decades of the twentieth century, making model organisms—new forms of laboratory life that scientists use as experimental tools, such as fruit flies, mice, and tobacco mosaic virus—became inextricable from the study of life.[31]

By the 1930s molecular biologists had come under the sway of the protein paradigm (which held that proteins governed heredity), a research agenda that lasted until the structure of the double helix was elucidated. Biologists then set about solving what they, in a Cold War informatic idiom, coined "the coding problem."[32] "Life," in these years, was a unitary quality common to all organisms. In 1954 Jacques Monod announced, "Anything found to be true of *E. coli* must also be true of elephants."[33] The locus of this underlying unity, for midcentury biologists, was molecular: organic molecules (proteins and nucleic acids, but also lipids and carbohydrates) were responsible for all living processes (such as respiration and reproduction). Life scientists' task was to determine analytically the fundamental mechanisms that join living structure to living function. For most of the twentieth century but reaching its height during the Cold War, scientists' approach to "life" was to treat it as something coherent and uniform, even simple. Its essence could be reduced to an elegant register of genetic instructions.[34] In 1967 molecular biologist and historian of science J. D. Bernal proclaimed, "Life is beginning to cease to be a mystery."[35]

The invention of recombinant DNA technology—at first also termed "synthetic biology"—in 1973 allowed scientists to ferry bits of DNA between organisms of different species, granting researchers an unprecedented amount of control over molecular biology. Biotechnologists could now modify organisms at the genetic level. For the last forty years, such recombinant organisms have been engineered under the aegis of biotechnology to fulfill human-defined needs (biomedical, pharmaceutical, agricultural) and to make money (e.g., bacteria that metabolize oil slicks, frost-resistant strawberries, mice riddled with tumors).

In the 1980s and 1990s biologists' fascination with genetic sequence and genetic data grew, culminating in the Big Science endeavors of the late twentieth century, the most famous and ambitious of which was the Human Genome Project.[36] By century's end, biology (like many other technoscientific fields in the United States at the time) was equipped with increasingly powerful computers and progressively abundant funding. The study of life became a campaign waged at the molecular level, one delirious for data.[37]

It is now commonplace for biologists to grouse that biology in the age of bioinformatics has become driven by data rather than theory. That is, biology in the first decades of the twenty-first century is in the depths of a "theory crisis." Rather than being animated by a coordinated practical inquiry into life toward which biologists bend their efforts, biology has recently become a discipline lacking a theory. We have arrived at a moment when many say that science is "data driven." Popular-science magazines such as *Wired* ring the death knell of science, announcing that we have reached "The End of Science: The quest for knowledge used to begin with grand theories. Now it begins with massive amounts of data. Welcome to the Petabyte Age." The title of this article in *Wired* declares: "The End of Theory: The Data Deluge Makes the Scientific Method Obsolete."[38] Such proclamations about the end of science and the end of theory recall Francis Fukuyama's myopic augury in 1989 that the fall of the Berlin Wall heralded the "end of history." History's demise was greatly exaggerated, and the same is true of science.[39] Perhaps the scientific enterprise is perpetually haunted by its own obsolescence. Biology is not in any way endangered— "theory crisis" is not meant to evoke hand-wringing anxiety but merely to identify a changing mode of biological reasoning.

Historians of science are now interrogating Big Data, suggesting that the age of Big Data commenced centuries before the government funded Big Science endeavors, databases, and digital computing in the mid-twentieth century, even before the origins of biology itself.[40] Historian of biology Bruno Strasser has argued that the life sciences have *always* been reliant on Big Data, and that contemporary biological databases are analogous to eighteenth-century natural history collections.[41] Collection and comparison may have long histories in the life sciences, and there is more to modern biology than hypothesis-driven experimentalism. Historians of science parse "data" and "theory" respectively as the *content* and *aim* of scientific databases—populate a database with enough data, and theories will emerge. Nonetheless, the meaning of these two terms in biology has radically transformed in recent decades.

Namely, "data" assumes *meaning*. It is necessarily information collected and classified in service of or as evidence to be analyzed to prove or disprove a hypothesis or to bolster a theory. And molecular biology is no longer guided by theory, whether hypothetical, experimental, or otherwise. Molecular biology, bioinformatics, and genomics swim in a torrent more noise than signal: genomes sequenced, protein structures uploaded, and most of the rest databased, tabulated, and released online. Moving ever outward to map out increasingly dense and complexly interrelated living systems, grasping after some explicandum for life, the life sciences are now overrun by proliferating -omes (genomes, to be sure, but also proteomes, interactomes, microbiomes, transcriptomes, epigenomes, regulomes, secretomes, foodomes . . .).[42]

Today, life is again an enigma, and not the sort pliant to cryptographic deciphering. Though biology remains for synthetic biologists (like mid-century molecular biologists before them) amenable to physicochemical and material analysis, the turning point presaged by what some now call "the postgenomic era" has practical consequences, not just for *which* living things now occupy our world but also for *how* some biologists understand what "life itself" is.

Put otherwise, and following Foucault's phrasing, if life did not precede the late eighteenth century, perhaps it did not survive the twentieth century. "Life" as an analytic object has come undone. Seeking answers, synthetic biologists build new living things, and in so doing they retroactively define what counts as "life" to accord with the living things they manufacture and account to be living. The organisms conceived by this latest crop of mechanical and electrical engineers-cum-biologists, then, are altogether different from the creatures built by biotechnologists: while some are made to serve discrete pharmaceutical or agricultural functions, many of them are made as a way of *theorizing the biological.* Rather than being the common denominator of all living things, "life" has once again become a *problem* of ontological limits and discontinuities. In 1954 what was true of *Escherichia coli* may have also been true of elephants. But in the twenty-first century, synthetic biologists seek to engineer both *Mycoplasma* and mammoths in order to figure out how both creatures uniquely morph and fragment formerly unitary definitions of life. As such, analyses of life are newly simultaneous with and enabled by synthesized instantiations of it.

To what end? *Synthetic* situates synthetic biology within this theory crisis precisely because synthetic biology is one of this crisis's immediate sequelae. Synthetic biologists arrived upon biology's "data deluge" and responded: if all the data collected, collated, coaxed, and tended by experimental inquiry are *not enough,* then biological manufacture will serve as life's new "theory machine."[43] Just as genetic *synthesis* and *sequencing* are terms of art for the technologies that drive synthetic biology, *synthesis* and *analysis* are joined philosophical modes of reasoning underwritten by these paired technologies. Making stuff—synthesis—became a mode of analysis, a way of theorizing the biological.

The coming pages offer a historically informed anthropological analysis of synthetic biology. Anthropological, because I am interested in how what people *know* is informed by what they *do,* and in particular by the exquisitely complex relays between artisanship and epistemology. Historical, because I seek to understand better the continuities between this contemporary scientific enterprise and prior efforts to control and study nature, and in particular the deep historical significance of "maker's knowledge," the powerful knowledge that can be forged only by making (rather than finding) something. Though synthetic biologists often claim that synthetic biology is newly new, the field (like other technoscientific "breakthroughs") rearranges, recycles, and modifies a cluster of forgotten social, scientific, and historical possibilities that thread their way through modernity: "The rupture shows itself to be also a repetition."[44]

I unearth these repetitions to interrogate synthetic biology's congruence of technical and discursive knowledge, showing how biological making and biological knowing traffic both ways. This convergence is recognizable to synthetic biologists, who explain what they are doing to themselves and to others as building living things in order to understand life better. When technical and epistemic knowledge of life converge, the objects of synthetic biology function as *persuasive objects*. They convince synthetic biologists that life is marked by the qualities—technical, substantive, and social—that they ascribe to it. That is, biological analysis either precedes or is simultaneous with biological manufacture. This sort of active, interested making is endemic to the sciences of our time.

Sequencing *Synthetic*: A Note on Ethnographic Form

Paging through the classic ethnographies that defined the early cultural anthropology oeuvre, one finds chapters arrayed to offer a synoptic and holistic view of a single culture. Ethnographers such as Alfred Radcliffe-Brown and Edward Evans-Pritchard proceeded from a society's cosmology to its ecology, politics, kinship systems, and ceremonial beliefs. Such monographs were meant to be exhaustive. They treat cultures as isolated and timeless, yet at risk of vanishing, as entities that anthropologists could preserve in textual fixative. Since anthropological repatriation and the "writing culture" turn of the 1980s,[45] anthropologists have espoused a more experimental, critical, eclectic, and reflexive approach to ethnographic writing.

In appropriating traditional ethnographic conventions in this book—progressing from religion (chapter 1) to kinship (chapter 2) to economy and property (chapter 3) to labor (chapter 4) to household (chapter 5) to origin tales (chapter 6)—I do not mean to move generically backward. Rather, I self-consciously appropriate canonical ethnographic conventions and objects of anthropological analysis in order to upend them, showing how scientists build such cultural categories simultaneously into objects of biology and into objects of theory. That is, these chapters do not report on synthetic biologists' kinship systems or economic transactions. Instead, I take these categories as *ways of thinking* that are the actors' own. I use them to evaluate how synthetic biologists build established and habitual forms of life (e.g., kinship, exchange, and religion) into putatively new synthetic life-forms. To that end, I demonstrate, for example, how a redesigned bacteriophage speaks to synthetic biologists' religious rhetorics, and how biocatalytic pathways usher in new ways of thinking about relatedness.

Chapter 1, "Life by Design," tracks early experimental efforts to model and engineer genetically simplified viruses at MIT in the early 2000s. I reveal the way Judeo-Christian tropes of creation shape synthetic biologists' descriptions of this work. By identifying themselves as both agents able to "evolve" life and as animals subject to evolution, synthetic biologists imagine themselves in epistemologically ambiguous territory when they redesign living systems.[46] Weaving evolutionary tales with biblical ones, they cast themselves as figures simultaneously unnatural, natural, and supernatural. I place this account within the context of arguments over creation and intelligent design that were then prominent in American political discourse in which both my interlocutors and I were steeped, using those debates to clarify how synthetic biologists were thinking about what it means to, as they put it, "intelligently design" life.

Chapter 2, "The Synthetic Kingdom," examines Bay Area synthetic biologists building organisms that manufacture fuels and drugs, such as Jay Keasling, whose goal was to manufacture cheap antimalarial drugs in bacterial hosts. I appraise how these researchers think about the microbes they made: *E. coli* and yeast, brewing in flasks, containing genes from disparate kingdoms and domains of life. Joining the history of biological taxonomy to anthropological theories of queer or "voluntary" kinship, I demonstrate that such organisms inaugurate new forms of relatedness, which synthetic biologists treat as both putatively "natural" and phylogenetically ambiguous. This chapter adds a new category, the "postnatural," to chapter 1's tripartite categorization of the natural, unnatural, and supernatural. Such organisms do not fit neatly into trees of life based on descent, ancestry, or lineage. Some scientists and artists even debate whether they are building a new branch of the tree of life, "Kingdom *Synthetica.*" Others emphasize how biological "relatedness," as they put it, has become newly transactional: rather than being gene-based, the hybrid organisms they make are connected by "rational choice." Their efforts undermine any notion of biological relatedness as fixed or natural. It may seem counterintuitive to use queer kinship theory to describe bioengineered organisms. I nonetheless contend that, just as queer kinship problematizes Euro-American faiths in blood-based nuclear relatedness, engineered life-forms behave queerly, undermining theories of descent and lineage even in those organisms that have not been transgenically tampered with.

Chapters 2 and 3 are paired. Cultural anthropologists have demonstrated that social norms governing relatedness extend to the allocation and distribution of goods. On the heels of exploring synthetic sorts of transgenic kinship, I turn to how such novel organisms mirror synthetic biologists' arguments about how such creatures should be exchanged. Chapter 3, "The Rebirth of the Author," extends this argument about new forms of transgenic kinship by showing that ideas about genetic exchanges within biotic systems underwrite the ways biologists think about economic exchange, in particular how analogies of life to either text or machines shape the way synthetic biologists think about and act upon intellectual property decisions ranging from patenting and copyright to credit, attribution, and publishing. Not only shaping epistemological definitions (of what life *is*) or normative claims (about what it *should be*), these analogies impact the way bioengineering gets done and the legal and economic regimes that are installed in and around it.

I next examine how synthetic biologists seek to standardize not only parts and economic exchanges but also labor practices. In Chapter 4, "Biotechnical Agnosticism," I enter the lab of a Boston start-up company that built what members term a biological "assembly line" following the principles of Taylorism, the late-nineteenth-century management theory that tried to maximize labor efficiency. I compare it to a much larger, for-profit synthetic biology company in the Bay Area, in which the corporate ethos is also suffused by management theories emphasizing efficiency. Both companies subscribe to the "Toyota Way" production cycle forged in Japanese factories and popularized in American manufacturing philosophies such as General Electric's "Six Sigma." Touring both labs and speaking to their founders, managers, and bench-workers, I observe the deskilling of PhD benchwork in favor of undergraduate labor in one company and short-term manual laborers operating robots in the other. Juxtaposing these two companies, I show how engineers have imported not only technical principles of manufacture (such as standardization and abstraction) into biology but also the labor relations and forms of alienation that underwrite mass production in late capitalism. As a result, synthetic biological work is fragmented, divided between the high-prestige work of biological design and the automated tasks of biological manufacture. Indeed, in some cases biological design is evacuated from industrial synthetic biology, which instead is feverishly spurred on by ever-increasing speeds and scales of production. The chapter ends by asking what becomes of the synthetic biologist when biological labor no longer needs her tending.

Extending this line of argument, chapter 5 offers one rejoinder to chapter 4's ironic upending of the convergence of biological making and knowing. "Life Makes Itself at Home" turns to new locales in which nonbiologists capitalize on the notion that biological engineering is becoming a "deskilled" task by choosing to conduct bioengineering in nonacademic locales. I observe how amateurs and hobbyists use the same genetic parts developed by synthetic biologists to engineer living systems outside professional laboratories: in kitchens, garages, and community hobby work-shops. I arrive at the conclusion that, in occupying ambiguous territory between experimentation and manufacture, synthetic biology has allowed bioengineering to leave the laboratory. Today dilettantes dabble in synthetic biology in order to argue against the large-scale, deskilled, and highly proprietary status of industrial synthetic biology (and "Big Biology" and biotechnology more broadly). Doing biological work at home, these biohackers subvert and critique this state of affairs, demonstrating that bioengineering may be a domestic enterprise.

Chapter 6, "Latter-Day Lazarus," inverts the ethnographic genre, ending rather than beginning with an origin story. If the first chapter interrogates how synthetic biologists animate theories of "good" biological design to cast themselves as creators of a more rational Eden, the final chapter responds, in good Judeo-Christian fashion, with a resurrection tale, one in which synthetic biologists propose engineering a bio-Eden out of a biological wasteland. Synthetic biologists and conservation biologists are jointly working to "revive" extinct species such as the woolly mammoth and the passenger pigeon. Reporting on their efforts, I pay attention to narratives of biologically enabled historical salvage, returning to the theme of trans-species and interspecies exchange from chapter 2, asking how synthetic biology problematizes definitions of species purity. By bookending "life" as an epistemic category, synthetic biologists, I conclude, remake biology as not just physical living stuff but also an engineered demonstration of what living things might once have been and what engineered organisms might someday become.

Given this book's title, I would be remiss if I were not to ask: what is the "synthetic"? How did it become a descriptor appended to the word "biology"? Toward answering these questions, I interleave each of the chapters with an interlude that meditates on earlier enunciations of "synthetic." Written as keyword entries, each interlude is loosely coupled to themes animating the following chapter. Because synthetic biology is more "repetition" than "rupture," the ways in which notions of the "synthetic" have traveled in contexts beyond synthetic biology shed light on its current usage in the life sciences.[47] They reveal embedded, forgotten, revived, and perhaps unintentional meanings of "synthetic biology," as well as the qualities, characteristics, and valences conveyed by contemporary researchers' efforts to achieve "synthetic life." Synthetic biology can productively be compared to prior examples of the *synthetic* understood as variously and equivocally referring to that which is unnatural, man-made, assembled, fake, artificial, imitative, deductive, constructive, composed, better than nature, and a danger to nature. Tracing the history of the term therefore uncovers some of the surprising ways in which ideas of synthetic life are indebted to Lycra spandex, margarine, the paintings of Georges Braque, and the music of David Bowie.

Each of these interludes traces prior enunciations of "synthetic," showing how such connotations infuse how synthetic biologists talk about the work they do. Etymological imports from logic, chemistry, and music are lively, burrowing into and embedding in the synthetic, animating synthetic biologists' projects and their thinking about life. These brief histories reveal how multiple meanings of "synthetic" inhere in the current usage of "synthetic biology," which makes the field's name productively ambiguous, polyvalent, and slippery. "Synthetic" capaciously refers simultaneously to a way of *reasoning* about biology (not analytic), a way of *conducting* biological research (not experimental), and the material *products* of that thinking and working ("synthetic life").

This is, then, an argument about circular reasoning: synthetic biologists build new living things to test theories, and the things they build function as materialized theories that in turn bolster ideas about what the biological *is*.[48] However, this thinking is not neatly circular, because it is done by people working in a particular time and place. Synthetic biologists are, needless to say, social and historical actors. As such, all sorts of political, economic, cultural, pop-cultural, and technological circumstances graft themselves into the looped logic by which making engenders knowing.

As a historical phenomenon, synthetic biology is indebted to the state of the life sciences—and molecular biology specifically—at the turn of the twenty-first century. However, synthetic biology is also part of a larger constellation of postmillennial cultural stuff. A partial list of events witnessed during the years about which I write includes the post-9/11 "War on Terror"; the aftermath of the dot-com bubble burst; the rise of the Maker movement, DIY (do-it-yourself), and hacker cultures; the Space Shuttle *Columbia* disaster; the growing popularity of copyleft, Open Source, free software, and allied movements; bombings in the subways of London and Moscow; the launch of online social media and digital platforms such as Facebook, Twitter, Wikipedia, and WikiLeaks; the Arab Spring; the advent of consumer 3-D printing; the resurgence and retrenchment of the American religious Far Right; the tragic regularity of ecological and urban environmental disaster (tsunamis, earthquakes, hurricanes, and floods), as well as more obviously anthropogenic catastrophes (the Fukushima-Daiichi nuclear meltdown, the Deepwater Horizon oil spill); the mainstreaming of the US LGBTQ movement; the global Great Recession. Newly built biological systems are *materialized theories*, but they are also postcards from a particular cultural moment.

I reparatively read synthetic biology as an effort, during these years, to nourish life-forms and forms of life that could unmask a naturalized and categorical understanding of what "life" is.[49] A queer appraisal of synthetic biology attends to which "postnatural" creatures are allowed to survive, thrive, and revive in our world (and which are not). Engineered organisms smuggle conventional ideas about creation, kinship, property, labor, democracy, and species beneath their immaculately ahistorical membranes. Yet the very existence of such "life otherwise" could nonetheless promise to reorient or displace habitual ways of acting upon and knowing the world.

At a time when biology is recognized as more complicated than molecular biologists ever could have anticipated, synthetic biologists respond by coaxing new creatures into recombinant life. These brave new organisms grow, mutate, metabolize, divide, and senesce, yet they also speak eloquently of their times, of nature and artifice, of analysis and synthesis, of life and its limits.

Plastic Fantastic

syn·thet·ic (sĭn-thĕt'ĭk)
adj.
(In most senses opposed to NATURAL *adj.*)

1. Of a substance; manmade (rather than objects fashioned from natural materials)
    a. Relating to that which is inauthentic, fake, or cheap;
        > 1962 RACHEL CARSON *Silent Spring* The chemicals to which life is asked to make its adjustment are no longer merely the calcium and silica and copper and all the rest of the minerals washed out of the rocks and carried in rivers to the sea; they are the synthetic creations of man's inventive mind, brewed in his laboratories, and having no counterparts in nature.
    b. That which exceeds or is superior to its natural counterpart: cf. SYNTHETIC 1A. *Obs. Rare.*
        > 1978 X-RAY SPEX *Day the World Turned Day-Glo* The X-rays were penetrating / Through the latex breeze / Synthetic fiber see-through leaves / Fell from the rayon trees.[1]

A rtificial silk, invented in 1885 during an outbreak of disease among silkworms in France and renamed rayon in 1924, is, strictly speaking, a "man-made" but not a "synthetic" fiber.[2] In the synthetic fiber industry, "synthetic" refers only to fabrics made using petrochemicals, rather than those that modify a plant or animal product. Rayon was derived and spun from plant cellulose.[3] As automobile manufacture expanded in the first decade of the twentieth century, the need for leather interiors outpaced leather tanning, a market gap that necessitated "artificial leather," first marketed under the brand name Fabrikoid (Leatherette, Corfam, and Naugahyde were some of the leather substitutes developed later). Such products were marketed as not only cheaper but *better than* the "real thing."[4]

Alongside American manufacture, the synthetic fabrics industry was developed in Italy to compensate for shortages of "natural fabrics" during

World War I and continued to grow during the subsequent economic decline of the 1930s. These fabrics were adopted as icons of futurism, with its technoaesthetic of speed and functionality: Italian textile manuals portrayed "rayon not as artificial but rather as an intensified, accelerated, redeemed prolongation of a (national) natural world that has been emancipated and democratized by modern science."[5] The Italian fashion and interior design industries accelerated the aestheticization of the synthetic, spinning shortages into market aspiration—as, for example, in Salvatore Ferragamo nylon platform shoes.[6] Here, the synthetic is not opposed to the natural but identified as nature in its heightened and distilled form—an object of desire, unmoored from the exigencies, rhythm, and scale of the natural world.[7]

"True" synthetic fabrics (i.e., those that are entirely chemical and laboratory manufactured) were pioneered by the DuPont Company, which dominated the American chemical industry throughout the twentieth century and was the world's leading producer of synthetic fabrics for forty years following World War II.[8] DuPont condensed its corporate profile in 1935 to the slogan "Better Things for Better Living . . . through Chemistry." In the decades of "commodity scientism,"[9] DuPont's fabrics received slick names befitting modernist aesthetics, the glossolalic nonsense poetry Don DeLillo invokes in *White Noise*: "Dacron, Orlon, Lycra Spandex."[10] Plastics such as Lucite and Butacite replaced household products made from ivory, metal, and glass.[11] Japanese embargoes on silk opened up the American market during World War II for nylon, which the corporation developed between 1929 and 1935, first marketed in 1940, and used in products from parachutes to lingerie.[12] Dacron, a synthetic fabric DuPont licensed and began selling in the 1950s, would by the 1960s be used to fabricate tubing for artificial hearts, after surgeon Michael DeBakey sewed a prototype on his wife's sewing machine.[13] Neoprene synthetic rubber (first marketed as "Duprene" in 1931), Orlon (a wool substitute; 1948), Cordura (1950), Mylar (1952), Tyvek (1961), Lycra spandex (1962), Kevlar (1965), Nomex (1967), and other "jet age" synthetic materials increasingly coated, wrapped, reinforced, glazed, and laminated Cold War American surfaces.[14] Chemical companies worked alongside the fashion industry in the 1960s to forge a "synergy of technology and taste,"[15] portraying synthetic fabrics as upmarket, progressive, and fashionable. According to one DuPont manager, chemists could make "a better fiber by design than a sheep produces inadvertently."[16] For midcentury Americans, synthetics were not substitutes for the real thing; they replaced and superseded it.[17]

FIGURE INT-I.I. "Dancing Molecules" wearing synthetic fabrics, from the Wonderful World of Chemistry show at the DuPont Company pavilion at the 1964–65 New York World's Fair. Courtesy of Hagley Museum and Library.

Yet in the 1960s ecologists and conservation biologists placed the synthetic under pressure, leading consumers to favor natural fabrics over their synthetic counterparts. In this period, ecologists called chemical companies to account for the dangers that chemically synthesized products might pose to the natural world. In her landmark book *Silent Spring*, Rachel Carson sounded a warning: "The chemicals to which life is asked to make its adjustment are no longer merely the calcium and silica and copper and all the rest of the minerals washed out of the rocks and carried in rivers to the sea; they are the synthetic creations of man's inventive mind, brewed in his laboratories, and having no counterparts in nature."[18] It is easy to read in the contempt in which many Americans today hold synthetic fibers both a disenchantment with technoscience and a great deal of class-based pretension.[19] After midcentury "miracles of science" failed to materialize—or, worse, proved to be ecological scourges—synthetic fabrics were viewed with suspicion: "Synthetic fibres are miracles of science whose very nickname, the 'man-mades,' draws attention to a kind of scientific power that is now widely questioned."[20] This sense of "synthetic" occupies the same semantic space as "artificial" or "fake," as well as "imitated" or "simulated."[21] The synthetic, at the end of the twentieth century, meant a *threat* to nature, novel, pervasive, dangerous, and fundamentally different in kind from naturally occurring substances.

Today, trends in American middle-class food consumption similarly value, for those who can afford them, natural over synthetic, organic over mass-produced, local and seasonal over standardized and scientifically maintained foods. Yet technologically enhanced food was not always spurned. Synthetic foods have followed a cultural trajectory parallel to that of synthetic fabrics. In the decades following World War II, "foods of the future" embodied a modernist, utopian, and futurist cachet (or "commodity scientism"). Meals-in-a-pill were predicted by World's Fairs and science fiction stories. By the mid-twentieth century, imitation foods joined synthetic fibers in simulating nature for better health and "better living."[22]

Historian Warren Belasco writes that from the contemporary vantage point of "an age when 'natural' and 'traditional' are far more appetizing food adjectives than 'synthetic' and 'artificial,' it is hard to understand how anyone could ever have awaited a future when air-conditioned, fully automated 'skyscraper farms' would raise algae on raw sewage in enclosed ponds and then pump the protein-rich green 'scum' to factories synthesizing cheap hamburgers and pasta."[23] Yet in the 1940s and 1950s "synthetic foods" promised, like the Green Revolution, a technical fix by which scientists might solve a global food crisis triggered by overpopulation. In the

1950s and 1960s saccharine and margarine were marketed not as cheap replacements for sugar and butter but as improvements on the real thing to be embraced by consumers seeking healthy lifestyles.[24] Where food was concerned, the synthetic once augured engineered solutions to social problems framed as natural ones—overpopulation, food shortages, overconsumption, and diseases of civilization.[25]

From its first usage to describe a human-made material artifact in 1874,[26] "synthetic" referred to manufactured materials or substances, especially those chemically synthesized. This use of "synthetic" gave it a gloss close to "man-made" or "imitation," as it came to describe products designed first to approximate or supplement natural materials, then to supplant and surpass them. As synthetic and organic chemists invented new products in the late nineteenth and early twentieth centuries, the list of things labeled "synthetic" grew: synthetic resins, indigo, silk, rubber. These products were marketed as better (or at least as good as), cheaper, more plentiful, or easier to procure than their natural counterparts. For a century, until their eventual decline in the 1960s after ecologists raised concerns that the synthetic chemical industry was devastating the environment, synthetics commanded a market share separate from the plant and animal products from which they were first derived.

While the antonyms of "synthetic," to our ears, might include "real," "authentic," or "genuine," "synthetic" has not always borne a negative valence of inauthenticity, fakery, or cheapness, nor necessarily does it for today's proponents of synthetic biology. They see their organisms not as cheap substitutes but rather as products of technoscientific ingenuity. Not necessarily "better things for better living," to borrow DuPont's phrase, but perhaps *better living things*.[27] For most of the twentieth century, the synthetic was artificial, but artifice was better than the real thing. Manufactured life, synthetic biologists claim, is arguably artificial, but it is undoubtedly an improvement upon nature and evolutionary caprice—life manufactured following human logic and design principles, for them, surpasses and refines any naturally occurring organism.

CHAPTER ONE

Life by Design: Evolution and Creation Tales in Synthetic Biology

"Them?"
"Nature and God."
"I thought you didn't believe in God," said Jimmy.
"I don't believe in Nature either," said Crake. "Or not with a capital N."
—Margaret Atwood, *Oryx and Crake*, 206

The perfect match, you and me
I adapt, contagious
You open up, say welcome
. . . .
The perfect match, you and I
You fail to resist
My crystalline charm
. . . .
My sweet adversary, ooh
My sweet adversary, oh
My sweet adversary
—Björk, "Virus," *Biophilia*

I had a virus I couldn't kick. Feverish and congested, I hurried from the MIT walk-in clinic to listen to Drew Endy lecture to the Department of Biological Engineering. On this overcast day in November 2005, Endy was one year into a tenure-track professorship at MIT (he had first arrived at MIT in 2002 as a research scientist). He looked young—about ten years younger than I knew he could possibly be, given his academic trajectory. The only aspect of his appearance that betrayed his age was his hair color, which had, in just the few months since the summer we had met, begun to fade from light brown to gray. He wore wire-rimmed glasses and kept his hair close-cropped, but photographs reveal that as a graduate student at Dartmouth in the mid-1990s, he had sported an abundant beard that suggested a previous incarnation as an outdoorsman. When not busy with teaching, researching, and preaching the gospel of synthetic biology, Endy blew off steam by whitewater rafting, hiking, and kiteboarding and would organize semiannual lab field trips to get his students off campus and outdoors.

Despite the intervening decade between earning his PhD and arriving at MIT, Endy continued to dress like a graduate student, a quirk that was tolerated, if not embraced, by the laid-back sartorial culture of MIT.[1] On any given day, Endy would wear a T-shirt advertising some aspect of his work in synthetic biology: shirts emblazoned with logos of the BioBricks Foundation, MIT, or Creative Commons, and one that merely promoted "DNA." Today was no different.[2] Perhaps some of Endy's persona was self-consciously constructed to incarnate a social type, even a caricature: the enthusiastic inventor, the youthful and magnetic leader of a new movement in scientific research. By 2005 he had become such a high-profile spokesperson for the field that his persona had already triggered backlash, with an editorial in *Science* noting obliquely: "Some of his peers privately complain that Endy is a larger-than-life self-promoter."[3] No wonder the room that day was packed, with students sitting cross-legged on the floor and overflowing into the hallway on the first floor of MIT's building 68, running along Ames Street.

After a superlative introduction by fellow professor Penny Chisolm, Endy launched into a lecture that was equal parts autobiography and research report: "So, to get started, this is how I got into molecular genetics and biology . . ." Over the next hour, Endy revealed that he too had a virus he couldn't kick. Indeed, he had been living with it for over a decade. This one, a bacteriophage named T7, didn't infect him (it feeds only on bacteria). It had, however, infected his thinking, spurring him to understand biology differently.

In this, as in many of Endy's talks, his style betrayed a tension between the logical and rigorous approach of an engineer, in which discipline he had trained, and the starry-eyed naïveté sometimes projected by scientists when presenting their work to a wider audience. Endy reported to the assembled faculty and students the origins of his current research: how, as a graduate student in the 1990s, he had developed a software model that, using data from sixty years of molecular biology research on bacteriophage T7, computed the complete intracellular developmental cycle of the bacteriophage, focusing on the infection of a single *E. coli* bacterium by a lone phage.[4] But his model, he told us, was lousy—it didn't work; it couldn't predict the behavior of T7.

Reflecting on his thwarted doctoral and postdoctoral work, Endy told his audience that those years were "pretty depressing to me, because now I'm coming back to this problem, where I want to *understand* how this thing works, and I want to *understand* how this thing works when I shuffle up all the [genetic] elements, right? And if I've got 72 elements, then I've got 72 factorial permutations, right? More than the number of protons in the universe, probably. And so it's not clear if I can build out a computer model that's going to let me explore this space, that I'm ever going to be able to get traction on this problem." Simmering down his question to a bullet point, he snapped, "What's *wrong* with the T7 genome?" I had been observing in Endy's lab for three months when I attended this lunchtime lecture, but his question shocked me nonetheless. I had never before heard a life scientist ask what was *wrong* with a living system. What sort of question was this—ontological? Normative? Ethical? Wrong to *whom*? Wrong by what metric? It was "wrong," I would learn, because it was disorganized and cluttered with genetic junk. It resisted Endy's best efforts to simulate, model, or understand it. The virus was wrong, in short, because it was a bad design. So he set about redesigning it.

Making Life Better

The MIT Synthetic Biology Working Group's self-described mission was to "mak[e] life better, one part at a time." The two labs constituting the group, led by Endy and Tom Knight, posted this slogan on their website when they founded the working group in fall 2002. If this book queries what synthetic biologists mean by "life," then this chapter draws upon ethnographic fieldwork among MIT synthetic biologists to ask what they mean by "better." Synthetic biology was—and remains—a diverse assemblage of interests, agendas, and research programs. Yet despite vast differences in academic background and wide variation in research agendas, these researchers are united by the philosophy that biology is a substrate amenable to the same engineering strategies employed by mechanical, electrical, and computer engineers to build the nonliving world, and they approach their engineering projects accordingly. Further, they are confident that building new living systems will advance their understanding of how biology works at a more fundamental and profound level than discovery-based experimental science can uncover: that manufacture will heighten understanding.

In this chapter I narrate the T7.1 project, one research agenda that dominated MIT's Synthetic Biology Working Group during my first years there. This was an effort to synthesize a "better" version of the genome of the T7 virus. In telling this story, I trace two lineages mirroring one another. First, I track how Drew Endy moved from a background in structural engineering into life sciences research, how he became the Principal Investigator of the lab in which I conducted much of my fieldwork, and how he reached the conviction that life can and must be understood by *simplifying it*. Second, I follow the career of a simple biological agent that drew Endy away from structural engineering and pushed him to think about questions of evolution and biological complexity. This humble bacteriophage (a virus that infects bacteria; literally, "bacteria-eater") both piqued Endy's curiosity and frustrated him. T7 is a bacteriophage that either replicates within or bursts bacteria (cycles scientists respectively call lysogenesis and lysis). Endy's encounter with T7—his "sweet adversary," to borrow a verse from Björk—encapsulates a constellation of concepts and terms that are central to MIT synthetic biologists' thinking about life and that will recur throughout this book: simplicity, minimalism, simulation, design and evolution, nature and artifice.

When these synthetic biologists set about to manufacture simpler forms of life, their thinking is animated by two altogether different understandings of "design." One construes their efforts as improving upon natural selection by "rationally" engineering living things in a goal-oriented manner. Such thinking, I show, is animated by a belief that evolution renders genomes that are cluttered, "junky," and poorly organized. The other takes design to be synonymous with "creation." As such, they imagine themselves to be both objects and agents of evolution.

In the early 2000s MIT synthetic biologists cast themselves in three very different roles. They simultaneously saw themselves as *unnatural*, building artificial organisms that are "fit" to thrive only in the artificial environment of the laboratory; as *natural*, doing the work that comes to them "naturally"; and as *supernatural*, effecting feats of biological engineering that render them divine. I arrive at the conclusion that these stories are animated by a religious discourse, which I evaluate using ethnographic examples culled from lab meetings, private conversations with graduate students, and published material. In invoking this language, MIT synthetic biologists slip between ideas about biological design, anxieties and hopes about "intelligent design," and Judeo-Christian accounts of creating life. Such stories cast MIT synthetic biologists as both godlike agents of biological evolution and unwitting participants in or targets of an evolutionary impulse.

Slouching Away from Bethlehem

How did Endy come to be delivering this lecture before MIT's Department of Biological Engineering? And what were the origins of his idée fixe with T7? Raised in Valley Forge, a small town in southeastern Pennsylvania, Endy, like many of the engineers with whom I spoke, remembered fondly youthful inclinations toward engineering, fueled by playing with Legos, Erector Sets, and Lincoln Logs.[5] Endy studied civil engineering at Lehigh University, a small college in Bethlehem, Pennsylvania, a postindustrial steel town less than two hours by car from his parents' home. The blast furnaces of the Bethlehem Steel Plant, now shuttered, still roared when Endy lived there, a symbol of American industrial manufacture. "The Steel," as it was called, forged iron for railroads, skyscrapers, and guns used during World War II.[6] Endy spent the summer of 1991 working for Amtrak, fixing bridges servicing the railroad between Washington, DC, and New York City. The shadow cast by Bethlehem Steel on Endy's early education struck me as especially formative when he explicitly compared—even denied any difference between—structural and biological engineering. As he rhetorically asked in his lecture, "What's the difference between building a bridge and designing a genome?" Such thinking denies any meaningful difference between the living and nonliving worlds, at least when it comes to their use as engineering substrates.

Bethlehem Steel is also the plant where Frederick Winslow Taylor first formulated his principles of scientific management,[7] a manufacturing philosophy some synthetic biologists also hope to build into biological engineering, making it faster, more streamlined, and less error prone by "standardizing" parts and protocols and setting up "assembly lines" for manufacturing engineered microbes.[8] After receiving his undergraduate degree, Endy remained in Bethlehem for two years to earn a master's degree in environmental engineering.

He next headed to Dartmouth, where he embarked upon a PhD in biochemical engineering and biotechnology. It is here that he first encountered T7. As he recalled for his audience, as a graduate student at

Dartmouth, his doctoral research involved building a software model that would simulate and predict the behavior of T7. Could biologists, he hypothesized, use fifty years' worth of experimental data to predict growth rates of viral plaques (infected bacterial cells grown in culture)?

To understand the stakes of this question, we must pause Endy's trajectory into the synthetic biology lab to trace the history of bacteriophage T7, asking how it too became an object of synthetic biology. Bacteriophages are some of the best-understood and well-characterized infectious agents in biology laboratories. Milislav Demerec and Ugo Fano, working at Cold Spring Harbor in 1944, are widely credited for isolating bacteriophage T7 from a standard anticoliphage mixture prepared by bacteriologist Ward J. MacNeal. T7 was the last virus isolated from a series of seven phages that were numbered in the order in which they were discovered (T for "type").[9] Experiments with T7 demonstrated in 1952, just one year before Watson and Crick elucidated the structure of the double helix, that bacteriophages were near-perfect parasites—they assimilated and converted all host DNA into viral DNA.[10] A few years later, researchers reflected on the role of RNA by studying the behavior of T7, concluding that it was "possible that the specific kind of RNA synthesized by the host under the influence of the infective phage may serve as the proper functional unit for the synthesis of phage specific protein."[11] This observation helped midcentury biologists to lay down the "Central Dogma," the tenet that "DNA makes RNA makes protein."

A slim genome, T7 was in 1983 one of the first living things to be sequenced, as it comprised fewer than forty thousand base pairs.[12] It was simple enough and short enough for its sequencing to be tractable by 1983's standards. Although T7 first snuck its way into molecular biology labs, smuggled within the bacterial Trojan horse it had infected and whose DNA it slowly converted into its own, by the time Endy began studying it as a graduate student fifty years later, it had become a workhorse of molecular biology and genetics, arguably one of the most comprehensively understood objects of biological experimentation. Hence, Endy was curious as to whether T7 could be modeled computationally—as he put it in his lecture, "whether or not our understanding of this relatively well studied natural biological system is good enough to support analysis."

In his 1997 doctoral thesis, Endy writes that his work was "motivated by the desire to develop the coupling between the information database and reductionist tools of the biologist and the synthetic tools of the engineer. . . . To improve our understanding of biological systems and through such understanding better apply them."[13] He used his programmed model to try to predict what would happen in mutant versions of the same virus. If you moved around chunks of viral genetic material called coding regions (the "seventy-two elements" Endy would mention in his lecture) to make viruses that, for example, expressed RNA polymerase in a different order than they had before, would the computer model still be predictive? Building a model, on his reasoning, should effectively verify the sum total of the data molecular biologists had gathered about bacteriophage T7 during its long tenure in research laboratories.

But this was not as simple a doctoral research project as Endy had hoped. It would stretch beyond his graduate work to animate his postdoctoral work (and later would be taken up by his graduate students at MIT). Splitting his postdoctoral work between the University of Wisconsin at Madison and the University of Texas at Austin, Endy began studying genetics and microbiology, thinking that because he had trained outside the life sciences, perhaps he had missed some crucial information about the virus in programming his model.

During these years, Endy compared the predicted growth rates of his computer-modeled mutant bacteriophages with the actual growth rates of the mutant viruses, which as a postdoc he modified and cultured in the lab. But the results, he found, did not square: the computational model was an awful predictor of actual viral growth. What, he asked us, does it mean when the sum total of molecular biology's published data on bacteriophage T7 fails to predict how the virus reacts to perturbations and modifications of its genetic material? Endy took it as a failure of experimental "discovery-based" biology: all the knowledge painstakingly gathered from classical genetics, molecular biology, and virology, everything life scientists had learned about T7 from 1944 to now, was *not enough* to predict the behavior of an infectious agent so simple it is arguably only marginally alive.

Because the software simulation of phage intracellular infection did not agree with the observed experimental reality of phage infection, Endy (and, later, his students) chose not to modify the rules and parameters of the simulation but to modify the genetic material of the bacteriophage itself, in hopes of building a virus simple enough that it could be modeled computationally. Let me repeat: because the model did not work, instead of scrapping the model, Endy decided to modify the *virus*.

In their history of objectivity, Peter Galison and Lorraine Daston suggest that what they call mechanical objectivity strove to achieve a naturalism so detailed that it mimetically approximated that which it was meant to represent, until this naturalism was superseded by trained judgment and artistry: "the whole project of nineteenth-century mechanically underwritten naturalism suddenly seemed deeply inadequate. For the image to be purely 'natural' was for it to become, ipso facto, as obscure as the nature it was supposed to depict: a nightmare reminiscent of Borges's too-lifelike map."[14] But Endy's work with T7 and its computational doppelgänger demonstrates not a striving toward naturalism, or even anything that might be termed objectivity, realism, or trained judgment. Rather, it betrays a lack of any real interest in "Nature . . . with a capital N."[15] Instead of the model explaining how the virus works, the virus now would explain how the model works.

For Endy and the synthetic biologists who would later continue to pursue this project, the collision of efforts to describe and to rebuild, to observe and to simulate, made life an unstable category—unable to understand how a living thing works, they remade it in order to render it more comprehensible. Historians and anthropologists of science have recognized how models materialize theories—they are both representations of scientific thinking and tools that guide research.[16] But in this case, the model functions not to guide the researcher in thinking about an external phenomenon "as it is" but as a blueprint with which to mold it "as it should be."

Debugging the Bugs

To understand what MIT synthetic biologists thought was "wrong" with T7, one must first ask what evolution is to a synthetic biologist. For those with whom I spoke and worked at MIT, it is both nature's greatest design principle and its worst flaw. Evolution, they say, makes living systems flexible and adaptive, yet it can also render them error prone, incomprehensible, and somewhat baroque, genetically speaking. Synthetic biology got under way at the end of the twentieth century and in the first years of the twenty-first, the same years that the Human Genome Project (HGP) wrapped up. HGP ways of thinking about the genome influenced synthetic biology's project.

Early detractors of the HGP complained that regions of noncoding or so-called "junk DNA" would be pointless to sequence; molecular biologists worried that tax dollars better spent on HIV research would be wasted sequencing genetic material that did not contribute to the human phenotype and that therefore lacked meaning.[17] Twenty-three professors (representing "virtual unanimity of the . . . faculty")[18] of the Department of Microbiology at Harvard Medical School published a letter urging their colleagues *not* to sequence the human genome.[19] In the letter, they ask "whether identifying the last nucleotide in a human genome really has deep scientific value" and question "how a complete sequence could be useful for understanding the organization of the huge human genome: the magnification is wrong, like viewing a painting through a microscope." They conclude that "to sequence the genome because it is there" is a pointless, even potentially fiscally disastrous, enterprise.[20] And once the human genome was sequenced, it produced something of an epistemic panic attack—*now what*? Even the coding regions of DNA were then and remain today, admit Francis Collins and Craig Venter (who competitively sequenced the human genome), of little experimental or technical use.[21] On the tenth anniversary of the day on which Venter and Collins joined hands with President Bill Clinton, who declared the sequenced genome "the most important, most wondrous map ever produced by humankind,"[22] the only positive outcome of the HGP that biologists and biotechnologists could agree upon was that sequencing technology had vastly improved—it was faster, cheaper, and under control.[23]

At both the start and the finish of the sequencing race, the genome was haunted by its own density, complexity, and overwhelming *meaninglessness*. And evolution was always to blame: junk DNA was assumed to be the accumulated detritus of evolution, every obsolete bit and bob of every preceding organism, every strange cul-de-sac on the wandering path toward "fitness," every scar of viral infection, piled up, stuffed in, and overflowing the genetic "code." Such thinking—of genomes as massive, incomprehensible, noisy, and buggy molecular ciphers—would make its way into MIT synthetic biology, along with a healthy dose of software developers' enthusiasm for debugging.[24]

Intentional Biology

After completing his postdoctoral work, Endy took a position in 1998 as a fellow of the Molecular Sciences Institute (TMSI) in downtown Berkeley, California. TMSI had been founded two years earlier by Sydney Brenner, a geneticist best known for his Nobel Prize–winning work on the roundworm *Caenorhabditis elegans.* He developed a fate map of this simple worm that tracked its cellular development and differentiation across its life cycle. Brenner would say of his worm, an organism he (and others) likened to a software program, "You can look at it and say 'that is all there is.'"[25] This statement could also be read as Endy's inspiration in its most distilled form: to be able to write a genetic code for T7 simple enough to be wholly transparent, legible, and predictable.

I would later hear Brenner lecture at the 2008 annual synthetic biology conference, held that year in Hong Kong. He was in town for a day, a brief stop between Japan and Singapore. He spoke of biology, the fantasy of its simplicity, the frustration of its evolved complexity, and the consequences of engineering new living systems. Unlike Endy, Brenner suggested that perhaps engineered organisms should make use of complexity, not tame it: "Complexity [in the animate world] has been achieved not by *design* but by a process of natural evolution achieved by selection. And I think we have to ask ourselves whether we want to give that up as an engineering principle. . . . Math is the art of the perfect, physics is the art of the optimal, and biology is the art of the satisfactory."

Brenner wanted to build an institute in which scientists could pursue their own research without the pressures of an academic career or the oversight of a private laboratory. With an initial infusion of $10 million in a five-year grant from the Philip Morris Company, he founded TMSI.[26] By 1998 geneticist Roger Brent had left a professorship at Harvard, frustrated by the way his academic position hindered his interest in annotating parts of the human genome as the HGP spat out reams of data. Brenner had left his post at the Scripps Research Institute, and the two converged upon downtown Berkeley, deciding to build TMSI in proximity to the University of California, Berkeley. Brenner stepped down as director in 2001, the same year that funding from Philip Morris dried up, and Brent took the reins. He would serve as president, CEO, and director of the independent and nonprofit institute until 2009. One of the topics around which TMSI scientists converged was "simplifying" and "quantifying" biology, a project the institute's mission statement termed "Intentional Biology." TMSI was a place where scientists from multiple disciplines—cell biology, mathematics, engineering—could trade ideas, rub shoulders, and collaborate on shared questions.

Endy would remain at TMSI as a research fellow for three years, leaving only to take up his position at MIT. While at TMSI, Endy and Brent began studying T7 together. Applying TMSI's principle of "intentional biology," they compared the growth rates of wild-type to modified phage, finding that the simulated models did not agree with their physical counterparts. As Endy recollected before the assembled audience at his MIT lecture in

November 2005, "The things we believe to be true, they go into the computer, and then we [did] the comparison . . . [and it] isn't lining up at all." In a paper they coauthored and published in *Nature*, Endy and Brent write that "in biological systems (and simulations), too much depends on chance interactions among small numbers of interacting molecules to yield behavior that is completely determined over time."[27] They posit that such biological complexity (which Endy would soon aim to eliminate via biological design) is to blame for rendering computational models inaccurate.[28]

Form and Function, Deformed and Reformed

When he arrived at MIT in 2002, Endy still had not settled his score with bacteriophage T7. Endy joined forces with Tom Knight to found the Synthetic Biology Working Group. Knight had already been at MIT for over forty years, having arrived in the early 1960s at the age of fourteen. With a graying Lincoln beard and eyes that suggest he is laughing at a joke that you are not in on, he describes himself as "your basic geek." Knight learned computer science in artificial intelligence researcher Marvin Minsky's lab at a time when immense computers used punch cards and batch processing. He made his name among the first generation of computer hackers by working on ARPANET and helping to develop the Lisp machine, one of the first single-user workstations. Turning in the 1990s to biology, Knight thought about living material as indistinguishable from computer code. One of his favorite catchphrases was "the [genetic] code is 3.6 billion years old; it's time for a rewrite." Such thinking infected Endy's ongoing struggle with T7.

Endy assigned the T7 project to his first graduate students and summarized in his 2005 lecture the last three years of work on the virus. The hypothesis he set out to test, as he put it, is whether "it's possible to produce an engineered surrogate genome encoding a viable organism whose behavior is easier to predict [than that of the wild-type genome]." Sriram "Sri" Kosuri, a doctoral student in biological engineering who had an undergraduate degree in biology from Berkeley, set to work redesigning T7 with Leon Chan, a graduate student in MIT's Department of Biology. Their project, which, following software nomenclature, they named T7.1, was funded by grants from the US Office of Naval Research, the Defense Advanced Research Projects Agency, and the National Institutes of Health. Leon Chan had already graduated when I arrived in the lab, but I learned much from Kosuri, a charmingly gregarious West Coast transplant with a shaggy head of hair and taste for lo-fi indie pop who was never too busy to sit down with me to talk about synthetic biology.

Over the course of several years, Chan and Kosuri used new sequencing technologies to begin to sort out some of T7's "clutter." Endy projected slides mapping out the viral genome of wild-type T7 alongside maps of the modified and streamlined genome. The first step that they took, he explained, was separating out overlapping genes that coded for separate proteins, so that genes could be manipulated independently of one another. The virus that had hitched a ride into molecular biology by multiplying inside *E. coli*, that had been one of the first semiliving agents to be sequenced, and that had lured and foiled Endy was being decluttered and rebooted.

In the paper reporting on their work, published two months before Endy's lecture, the questions the authors posed echoed those now asked by Endy: "should we also expect that the 'design' of an evolved organism would be further optimized for the purposes of human understanding and interaction? Evidence drawn from fields outside biology suggests that the answer is no."[29] As evidence, they cite the fact that T7 bears fifty-seven genes coding for sixty proteins, only thirty-five of which have any known function. Perhaps, they asked, could we "safely ignore" the remainder?[30] Note that the word "design" appears in quotation marks while "evolved" does not. "Designing" a virus still functions as a metaphor for the authors, while "evolving" hardware or software using algorithms (a technique they describe in the next paragraph) has sunk into generalized common sense, losing its scare quotes. Evolution is not limited to biotic media, nor is it understood as being a way of describing the relationship between populations of living things in a dynamic environment. Instead, it here simply means modification over time.

The call to "ignore" anything about the viral genome that is incomprehensible is not unprecedented in biology. Certainly, the form and function of life-forms have, for life scientists and philosophers, oftentimes paralleled social, historical, and political forms of life. Hence, the impulse to purge variation and embellishments from living things is tethered to a modernism that simultaneously seeks to eradicate the foibles of human history. For example, an 1830 debate between zoologists George Cuvier and Etienne Geoffroy Saint-Hilaire over the anatomy of mollusks bled into contemporaneous debates among French architects about design, social reform, and the future of urban society. Proponents of "architectural rationalism" proposed that design types, following anatomical types, should be the "basis for a utilitarian approach to architectural design, offering a modular method that was 'scientific' in its ahistorical extraction of a built object from contextual considerations."[31]

A century later, as Peter Galison reconstructs, a similar endeavor was under way in Austria. The logical positivism of the Vienna Circle infused the design principles of the Bauhaus in the interwar period. In this regard, "transparent construction" was a rationalist, functionalist "modern 'form of life'" espoused by philosophers and architects alike. In both philosophy and design, Galison demonstrates, logical positivists and *Bauhäusler* sought an "elimination of the superfluous" by logically assembling theories and principles out of simpler elemental units, whether of perception, color, or geometrical shapes.[32]

Synthetic biologists also seek to excise historical exigencies from T7— not social or political history but evolutionary history. Despite similarities in their modernist aesthetics to their precursors in the 1830s and 1930s— designing things undecorated, modular, transparent, simplified, and rationally arranged—they conceive of the relation of function to aesthetics in reverse. If Bauhaus designers like Gropius understood functionality as arising from aesthetics,[33] here simplification and the elimination of evolutionary "ornament" promote biological utility, which MIT synthetic biologists equate with comprehensibility. Biological simplicity is aesthetically pleasing *because* it is transparently intelligible, rather than vice versa.

To drive home his point that good genetic design should lend itself to human comprehensibility, Endy next compared T7 to two different electrical circuits with the same function: both circuits take the square root of an input voltage. An engineer designed the first circuit. The second, like T7, is an "evolved artifact." A research group headed by John Koza at Stanford University ran a series of simulated evolutionary algorithms on a computer to "evolve" a new design for an electrical circuit. Such "genetic programming" (which, unlike actual evolution, is teleological) begins with a list of criteria for an end product. A program generates legions of possible designs, then allows the more robust (or "fit") designs to continue to "mutate" toward some desired end.[34] Endy reported that he had shown both circuit designs—the designed and the evolved—to fellow MIT professor Gerald Sussman, a researcher whose work focuses on automating scientific research and reasoning.[35]

Endy said that Sussman "refuses to explain to me how this [second circuit] works. It's not that he couldn't figure it out, but it *isn't designed for it to be easy for him to figure it out*, and so he chooses not to do it." The first circuit, however, is "easier to understand; if you wanted to change it, it would be easier to change. . . . And so now if you come back to this representation of an evolved piece of DNA, is it optimized for purposes of human understanding? . . . The hypothesis we [in the lab] got interested in testing was the answer might be 'no.'" Endy elicited laughs from his audience when he said, "we're all familiar with Darwin and the idea that evolution is cool, and it is. You know, but there's the other view of evolution— that it's a tyrant, giving us mutation without representation." By this he meant that nucleic acids, unlike software code, are not annotated to provide clear signposts or instructions for future engineers or programmers. The problem with evolution, on this view, is that while it may (putatively) maximize functionality, it is not easily legible on a genetic level. Such reasoning privileges genotype over phenotype, going so far as to erase phenotypic characteristics as criteria by which to assess evolutionary adaptation.

Similarly arguing by reference to Koza's electrical circuits, Chan and Kosuri write in their published paper that "so-evolved systems lack human readable descriptions and are difficult to understand, fix, and modify for new applications. By contrast, a structured design process produces systems that, in addition to functioning, are designed to be *easy to understand* and extend."[36] They explicitly based their approach not on the prior work of genetic engineers but on the practices of computer engineers and software designers, declaring that they were "inspired by the practice of 'refactoring,' a technique typically used to improve the design of legacy computer software."[37]

Endy, Chan, and Kosuri's overarching comparisons of T7.1 to John Koza's evolved electronic circuits complicate the relation of design to evolution. For Koza, the application of evolutionary algorithms to design problems is a means of seeking elegant solutions to difficult engineering problems (with mixed results). Here, evolution is a designer whose creativity is supposed to surpass human ingenuity, not the other way around. Yet Endy and his students treated natural selection as *opposed* to design— nature does not optimize, improve, or otherwise maximize itself, they were saying, but merely randomly accumulates modifications that may or may not be functional to the organism or "rational" to the engineer.

On Bad Design

Endy's description of phages as "evolved artifacts" rather than designed ones and his explicit alignment of a virus with an electrical circuit erase the distinction between viruses and circuits, taking them both to be "evolved artifacts." In so doing, he rejects any meaningful distinction between the organic and the inorganic. But more to the point, the parallelism asks how, if at all, evolution is related to design, as well as reanimates dusty ruminations on the relationship of organisms to machines. Thinking about living things mechanistically—imagining that organisms operate like machines, or that living systems, organs, limbs, and tissues function like machine parts—historically led thinkers such as William Paley and others to assume a Mechanic and to understand design as both the fitting of form to function and the fitting of means to ends.[38]

Design in nature has been a long-standing concern in biology, predating the discipline by several millennia, and many philosophers have cut their teeth on evidence of design found in nature. Evolution and design have not always been treated as mutually opposed. From the ancient Greeks to Thomas Aquinas, philosophers believed the natural world to be both ordered and purposive. Immanuel Kant, in providing an early definition of "organism," differentiated between organization and self-organization, specifying living things according to their internal or self-possessed organization. The difference, on this view, between organisms and artifacts is that artifacts are actually designed, whereas organisms can be understood only metaphorically "*as if* designed."[39]

English philosopher and theologian William Paley argued in *Natural Theology* (1802) that "the adaptation of each species to its environment indicated that it was designed by a benevolent Creator."[40] Elaborating on the analogy of God to a "divine watchmaker," Paley posited that merely observing a watch (and comparing that watch to a stone) made clear that "the watch must have had a maker; that there must have existed, at some time, and at some place or other, an artificer or artificers who formed it for the purpose which we find it actually to answer; who comprehended its construction, and designed its use."[41] Design, for natural philosophers, was evidence of a divine hand shaping nature and assembling parts such that the whole would function purposively.

Indeed, natural theologians of Paley's day, in thinking of body parts as mechanical devices, were led to theorize "God as an engineer."[42] And by 2005 the engineers reversed the analogy when they began casting themselves as godlike artificers. The question of design in nature has paired investigations into life's form with faith in Providence's role in molding that form. Thinking about life as already artificial—as either analogous to machines or indistinguishable from them—makes such analogies tick. Watches presuppose watchmakers, and the watchmaker is always divine.

Questions of perfection and error, of good design and bad design, haunt this discourse. Charles Darwin worried terribly about eyes, which, historian Jessica Riskin reminds us, "philosophers and physiologists from Aristotle and Galen onward had considered . . . to represent divine craftsmanship."[43] Here, the suiting of means to ends, the complexity of a living system, and the analogy of the eye to a mechanical device (often, the telescope) all suggested a divine designer rather than natural selection, a force that lacks teleology and intentionality. Yet Darwin was torn when it came to questions of design—sometimes, useless features constituted a design flaw, yet at other times, they were beneficial. In 1847 he noted that "all allusion to superintending providence [is] unnecessary. . . . [R]ather, expressly mention the design displayed in retaining useless organs for further modifications as proof of supervisal."[44] If natural theologians imagined a designer who crafted each organ and organism to a specific purpose and niche, here Darwin takes the "designer" to be one who sets good design principles in motion, including (especially) the retention of "useless" parts that might serve some unforeseen future purpose.

The question of "what is wrong" with a living thing would not have been thinkable or articulable to Darwin and his contemporaries—even "useless" parts must serve some later, as yet undefined, purpose. Such thinking also raises the specter of "intelligent design" theories, a term that in fact predates *On the Origin of Species* by at least nine years, when political philosopher Patrick Edward Dove wrote a treatise on intelligent design that questioned the use of both the terms "design" and "designer."[45]

Much more recently, theoretical biologists have taken up notions of biological design in opposition to neo-Darwinian theories, as, for example, in the work of Stuart Kauffman, who claimed that "organisms are *ad hoc* solutions to design problems [and] the answers lie in the specific details wrought by ceaseless selection."[46] Twentieth-century evolutionary biologists have also taken up the question of good and bad design, arguing over the place of bad design in nature, and what bad designs suggest about teleology, purpose, and the force of evolution. One of the most widely read examples of such thinking is the work of Stephen Jay Gould, who took an impish pleasure in pointing out just how nonoptimal nature could be, as evidence that evolution lacked any intelligent operator. His widely read essay "The Panda's Peculiar Thumb" argued precisely against intelligent design by pointing out that the panda makes do with an inefficient and awkward thumb-like protrusion to strip its bamboo—any reasonable celestial engineer would have built something much more suitable.

T7.1

MIT synthetic biologists, as engineers bending their design principles to living form, set about to do just that—to build biological systems more suitable, and in particular more suited to human understanding. To return to Endy's initial question: what is *wrong* with the T7 genome? The question suggests that something was made right by separating overlapping genetic elements. But what? As Endy showed the audience photographs of bacterial plaques infected by phage that had been grown in his laboratory, he explained that the reengineered virus, when transfected into *E. coli* and plated on petri dishes, resulted in reduced plaques. T7.1 was *much worse* than its predecessor at doing exactly what phage has evolved to do—infect bacteria, copy itself, burst open its host, infect more bacteria, repeat. The photographed plated cultures looked puny and sparse compared with the robust cultures of wild-type T7 that Kosuri and Chan had used as experimental controls.

If evolution is understood as the modification of an organism to be better suited to its own environment, then what characterizes evolutionary fitness or "good design" when "better" genomes make worse viruses? When MIT synthetic biologists bring their design principles to bear on living organisms, when they modify living things to conform to ideas about form and function imported from engineering disciplines, then "fitness" is also placed under pressure and defined recursively, according to the qualities MIT synthetic biologists define as fit. Synthetic biologists, by their own definition, are both part of the environment to which the phage must adapt and also outside or beyond it, demiurges dictating which organisms will succeed and which will fail.[47]

If a fit organism thrives in its environment, then the environment for which T7.1 is designed is an MIT laboratory in the Koch Biology Building. MIT synthetic biologists thereby insert themselves into evolution, making themselves the arbiters of which organisms are most "fit." As he flashed the slide of the photographed plaques on the overhead projector, Endy told the assembled faculty and graduate students that he will "ignore everything associated with the natural living world, and define an artificial living world that I completely *control* in the lab. . . . And this is it." Comprehensibility is an adaptive trait for phage in synthetic biology labs.

Five months later, I had dinner with Endy and Austin Che (a graduate student working with both Endy and Knight) in a Cambridge restaurant in MIT's neighboring Technology Square. Che is a quiet yet sharp-eyed man with a brutally sardonic wit. He arrived at MIT after completing an undergraduate degree in computer science at Stanford, and he retrained with Tom Knight toward a master's thesis that incorporated biology into his computer science training. The night of the dinner, he was two years into his PhD research. Conversation again turned to the viral plaques languishing in Endy's laboratory. Between appetizers and the main course, Endy posed the problem to me as a question: "Is there an environment that would naturally give rise to T7.1?" The question felt like a test. It tangled with definitions of natural and artificial, evolution and design: the use of the verb "give rise" effectively erased the years of graduate student labor that went into synthesizing and rebuilding the viral genome. Che smirked; he clearly had heard Endy pose this question before.

After a beat, in which I thought back to his lecture the previous November, I answered, "Yes, it's called a synthetic biology lab," and Endy laughed approvingly: "That's exactly what I say when people pose the same question to me." As he put it in his lecture when a grad student interrupted him to ask the same question, "So in nature, right, this [wild-type] one probably is going to rock. . . . In my lab, this [chimeric] one's going to survive, and this [other] one isn't." Inserting themselves into evolutionary narratives, synthetic biologists here serve as both agents and participants in a grand evolutionary narrative, one that artificially selects on the basis not of adaptability but of comprehensibility.[48]

The "Other" Intelligent Design

A year after Endy's lecture, at 5 p.m. on a Wednesday I sat in my usual spot in the back of the reading room on the fifth floor of the old Koch building at MIT, sandwiched between two bookshelves overfilled with back issues of *Science*, *Nature*, and *Cell*. I was by then acclimated to the mood of Endy lab meetings, which did not stand on ceremony. Grad students streamed into the reading room in pairs from the open adjacent laboratory door, midconversation, toting laptops and snacks. The room had a relaxed and convivial air, more so than other biology and biological engineering labs I had observed at MIT. Each week, a different student chaired the meeting. Today, Jason Kelly had donned the hat and held the scepter that marked him as presiding over the next two hours of research reports. Each week a different student was also tasked with providing snacks; Ilya Sytchev, a computer programmer developing a semantic web ontology for standard-ized biological parts, arrived a little after 5 p.m. laden with home-cooked foods prepared by his mother—salmon pirok, apple tarts, and fried dough shaped like roosters. Graduate students and undergraduate researchers appreciatively filled their paper plates while the lab technician made brief announcements. She lured students to upcoming safety training by telling gruesome stories of centrifuge rotors failing.

Endy arrived fifteen minutes late. He was giddy with news about a lec-ture he had just attended, in which he described seeing "kickass pictures" of magnetotactic bacteria.[49] He dubbed the research to the assembled graduate and undergraduate students as "just *soooo* fucking cool" and brainstormed how magnetotaxis could be implemented to serve functional ends in engineered organisms. Students next discussed whether after the lab meeting they should adjourn to watch a movie in the lounge down the hall, where a broken centrifuge was well stocked with beer. Chalk it up to disciplinary dispositions—feeling themselves to be at the forefront of a new approach to biological engineering, members of the Synthetic Biology Working Group cast themselves as firebrands and subversives, and one of the first things they did was dispose of many of the formalities and hier-archies of academic biological research. Over the next hour, as students reported on their week's work, they cracked jokes and interrupted one another. They believed they were on the cusp of something important, and their enthusiasm was infectious.

Imagine my surprise, then, when the atmosphere suddenly turned for-mal, even icy. Halfway through the meeting, a first-year graduate student in Computational and Systems Biology anxiously delivered his first pre-sentation to the lab. Despite clearly having carefully prepared his slides and talking points, he stumbled over his words and was red-faced and un-comfortable at the front of the room. He made the unfortunate mistake of beginning a sentence: "What Drew did after he created this [bacteriophage T7.1] . . ." The room, which moments earlier had been abuzz with ques-tions and side conversations, fell abruptly silent. Seconds passed. Endy spoke first, in measured tones. "We don't *create* biology; we *construct* it." The student blushed deeply from his shirt collar to his hairline, stammer-ing after Endy, "Umm, right, right, *constructed.*"

In the ensuing silence, I sympathized with the grad student, having ear-lier stumbled into a similar semantic snafu in an e-mail exchange with Reshma Shetty, a graduate student figuring out how to use DNA promot-ers and terminators as biological equivalents of logic gates in electronic circuitry. Shetty was a grounded and serious grad student who arrived at MIT to study with Knight after she finished a degree from the University of Utah. We met a few weeks after I began studying in the lab, as she had been traveling that summer introducing synthetic biology to undergraduate students in South Asia. After observing an undergraduate bioengineering teaching lab in which she served as teaching assistant, I had e-mailed her to double-check my understanding of the science behind the lab protocol. In my e-mail, I wrote, "The enzymatic activity of beta-galactosidase creates the pigment in those cells not exposed to light." In her response, Shetty redressed my language: "We like to avoid using the word 'create' in synthetic biology because of its god-like connotations and because it is not scientifically accurate. I would say that beta-galactosidase *generates* the pigment from a substrate instead of *creates*."[50] Clearly the callow graduate student and I had hit the same nerve, but why, I wondered, was it so important to synthetic biologists to describe their work as "constructing" rather than "creating"? The trouble, I realized, had by then been brewing for over a year.

While natural philosophers, Darwinians, and evo-devo (evolutionary developmental) biologists regularly invoked the divine in biological design, they were certainly not the only ones to do so. Religious proponents of creationism and intelligent design and supporters of evolutionary theory waged contests in American courtrooms, school boards, and congressional committees throughout the twentieth century. The Scopes trial upheld Tennessee's Butler Act in 1925, declaring it illegal for state-funded schools to teach evolutionary theory. Dozens of court cases have made similar rulings in other states and school districts. While creation science (a term coined in 1970 to rebrand "flood geology") holds that Earth is six thousand years old and that God created Earth and all life on it in six days, intelligent design (ID) proponents widely believe that while Earth is ancient, the complexity of life establishes the presence of a divine agent and planner.

Though the US Supreme Court ruled in 1987's *Edwards v. Aguillard* that teaching creationism was unconstitutional, it left open the possibility of teaching alternative scientific theories, and so intelligent design advocates repackaged intelligent design as a scientific theory. Republican politicians in the 1980s and 1990s, including Ronald Reagan in his presidential run, promoted teaching creationism and ID. A 1993 Gallup poll reported that 58 percent of Americans supported teaching creationism in public schools, while only 11 percent of Americans believed in evolutionary theory.[51] This fraught history suffused the way MIT synthetic biologists talked about creating and designing life.

In 2005 and 2006 ID had begun to inflect and infect synthetic biologists' thinking about their own design projects. MIT graduate students
described synthetic biology as "the other intelligent design." Shetty proposed in all seriousness at a time when synthetic biology lacked a dedicated peer-reviewed journal that such a journal be titled the *Journal of
Intelligent Design.* Thinking about synthetic biology as "the other intelligent design" put synthetic biologists in the awkward, if to them sometimes
flattering, position of having to think of themselves as life's Mechanics and
Watchmakers.

In December 2005 the US District Court for the Middle District of
Pennsylvania ruled in *Kitzmiller v. Dover Area School District* that teaching ID in US public schools violates the First Amendment of the Constitution. A volley of e-mails lit up the synthetic biology LISTSERV that afternoon and the next day. Tom Knight forwarded the 139-page court opinion
to the entire list, and Endy praised the language of the judge's decision.
Che responded more circumspectly, asking whether a defeat for ID was
indeed a victory for synthetic biologists: "Aren't we trying to show that
it is possible to intelligently design the biological world? Will we ever see
synthetic biology be used as evidence for intelligent design?" He ended
his e-mail by suggesting that if synthetic biologists were successful in their
project, they might unwittingly lend credence to creationists.[52]

Che, noticing that my anthropological interest was piqued whenever
synthetic biologists brought up design, had started forwarding articles to
me. He found one that had appeared on a pro-creationism blog in November 2005, the same month that synthetic biology landed on the cover
of *Nature* with the provocative line "life is what we make it." In the blog
post, an ID advocate drew analogies between synthetic biology and Paley's "old Divine Watchmaker."[53]

The author cited an article published by David Sprinzak and Michael Elowitz, synthetic biologists at Caltech who built a series of genetic
"switches" in bacteria. These "switches," when combined to form a "circuit," caused protein expression in the bacteria to oscillate cyclically, so that
the bacteria fluoresce either yellow or green. Sprinzak and Elowitz describe
their engineered system as a "synthetic genetic clock" and conclude their
article poetically: "perhaps at this stage one can learn more by putting together a simple, if inaccurate pendulum clock than one can by disassem-
bling the finest Swiss timepiece."[54] Again, life, for these authors, is best understood not by its deconstruction and decomposition but by its assembly.

The ID blogger quoted this particular paper at length, taking this pas-
sage as one of many examples of a "design theme" that is "ubiquitous [in

the *Nature* special issue], while references to evolution were merely assumed and seemed forced." What the blogger did not realize was that Endy
and his students took design to be a better approach to organizing living
systems than evolution, which they parsed as *unintelligently* designed—
hardly evidence of the Godhead. Remember, for example, Endy's comparison of evolution to a "tyrant." He used similar language in an interview
with the *Bulletin of Atomic Scientists*, in which he grumbled, "Intelligent
design, from an engineer's perspective, would have documentation, and we
don't see that."[55] By this he meant that an "intelligent" approach, from the
standpoint of an engineer, would require that every artifact come with its
own user's manual explaining how it works, which is notably absent from
living organisms—T7 did not evolve toward human comprehensibility. ID
proponents, on the other hand, read synthetic biologists' use of "design"
altogether differently—they take it not as a rejection of evolution as aesthetically and practically insufficient but rather as a refutation of evolution
as a mechanism of biological change in the first place.[56]

Bruno Latour notes the ubiquity of "design" in contemporary life, as it
"has been extended from the details of daily objects to cities, landscapes,
nations, cultures, bodies, genes, and . . . to nature itself—which is in great
need of being re-designed."[57] Asking what design now means, Latour
claims that "designing" holds a middle ground between revolutionizing
and modernizing. Designing is never creating, nor is it re-creating—it is
not "construction, creation or . . . fabrication"[58]—because design always
entails *redesign*. As such, "To design is never to create *ex nihilo*. It is amus-
ing that creationists in America use the word 'intelligent design' as a rough
substitute for 'God the Creator.' They don't seem to realize the tremen-
dous abyss that exists between creating and designing."[59]

In line with Latour's insight about the gulf between design and creation, synthetic biologists sometimes use "design" to denote engineering
work operating halfway between revolution and modernization—that is,
improvement through redesign. Yet at other times, they use it precisely
because of the creative semantic ambiguity in "intelligent design." It allows them to slip between thinking of themselves as designers and as creators, despite the distance between design and creation.

"We Shall Be as Gods"

At its origins, members of the Synthetic Biology Working Group compared themselves to God. In a 2002 seminar at MIT's Computer Science

and Artificial Intelligence Laboratory (CSAIL), Che posed the question, "Did God create us so we could become God ourselves?" Yet by 2006, synthetic biology had started coming under fire—from the popular press, citizen action groups, and other scientists—who accused synthetic biologists of wanting to "play God." No wonder everyone in the working group had begun shying away from using the word "create." A 2007 *Nature* editorial quoted a representative of ETC Group, a Canadian civil action organization that publicly critiques synthetic biology, as saying, "For the first time, God has competition." The *Nature* author even described the quotation as "justif[ied]."[60] The verb "create" had turned inflammatory, even as it added to the field's hype by suggesting that synthetic biology was indeed powerful enough to give God a run for his money.[61]

Further, MIT synthetic biologists leveraged the create/construct distinction to distance what they were doing from the "synthetic genomics" of Craig Venter, the *enfant terrible* of the HGP, who had lately begun work in the J. Craig Venter Institute (JCVI) to build an entirely synthetic organism (i.e., a cell whose genome was manufactured using DNA synthesis).[62] In response to a question posted on the synthetic biology LISTSERV about the semantic difference between synthetic biology and synthetic genomics, Endy replied: "synthetic biology = let's make biology easy to engineer + understand how the natural living world works along the way // synthetic genomics = let's construct genomes + talk about playing god." When an article appeared in *Newsweek* (June 3, 2007) about Venter's efforts to engineer a "synthetic" cell, the cover portrayed Venter looking presciently into the distance, his face illuminated in an ethereal glow. The headline announced: "Playing God: How Scientists Are Creating Life Forms or 'Biodevices' That Could Change the World." By all appearances, God, it turned out, was indeed a powerful white man with a beard. Tom Knight brushed off the press as merely "Venter playing god again." The comparisons of Venter to a latter-day in vitro God would only multiply when in 2010 he announced he had engineered an "entirely synthetic" bacterium. A *Time* article reporting on this feat waxed biblical: "In the beginning, Craig Venter 'created life' in a lab."[63]

The watchdog groups and protests against synthetic biology also focused on life's manufacture as edging dangerously close to God's territory. Journalists often ask Venter and his colleagues at the JCVI whether building synthetic organisms is tantamount to "playing god," to which research scientist Hamilton "Ham" Smith regularly responds, "We don't play."[64] In 2007 ETC Group issued a press release that nicknamed one of Venter's synthetic organisms the "Original Syn." One of the ETC Group's self-published reports on synthetic biology, which they distributed at a bioengineering conference I attended, bore a cover depicting a modified version of Michelangelo's *Creation of Adam*, in which Adam's hand is holding a Lego, which he is placing atop an assembled Lego model of the double helix (fig. 1.1). An article in the *Berkeley Science Review* reported on the work of Jay Keasling and the Synthetic Biology Department of the Lawrence Berkeley National Laboratory (titled "Intelligent Design: Playing with the Building Blocks of Biology"). It was similarly illustrated with color reproductions of Michelangelo's frescoes from the Sistine Chapel, including *Fall of Man* and *Expulsion from the Garden of Eden*. Half a page was devoted to another modified version of *Creation of Adam*, this time depicting God stretching out his hand to bestow a pipette on Adam (fig. 1.2).[65]

Such iconography functions on several levels. It renders ambiguous which figure represents the synthetic biologist: is it the Godhead or the human? And if the human, then are synthetic biologists beneficiaries of godlike powers, symbolized by pipettes and Lego double helices, or are they about to pay the price of their overreach? In these images, stories about creation and knowledge are entangled with warnings about hubris, impulse, and human recklessness. Origin stories are always also morality tales.[66]

While Endy made light of such controversies by pinning to the wall of his office an illustration of Adam and Eve standing nude before the Tree of Knowledge, others were more eager to distance the Synthetic Biology Working Group from any allusions to "playing God," whether leveled by fellow researchers, protesters, or journalists. Toward that end, the working group commissioned Laurie Zoloth, a bioethicist from Northwestern University, to comment upon and analyze the ethical issues raised by synthetic biology. In preparation for the second annual meeting on synthetic biology, she forwarded a list of questions to researchers at MIT who had convened a working group to hash out such issues. Presenting her thoughts to the "Synthetic Society" assembled in an MIT conference room in 2006, she offered one bullet point for our further discussion, titled "We will be as gods." The quotation references the book of Genesis, in which the serpent cajoles Eve to eat the fruit of the Tree of Knowledge: "on the day ye eat thereof, then your eyes shall be opened, and ye shall be as gods, knowing good and evil."[67] If synthetic biologists were worrying over the theological implications of their work, they were doing so from squarely within a Judeo-Christian tradition (although many of the graduate students at MIT

FIGURE I.I. Cover image of ETC Group, "Extreme Genetic Engineering: An Introduction to Synthetic Biology," January 2007. Courtesy of ETC Group and Reymond Pagé.

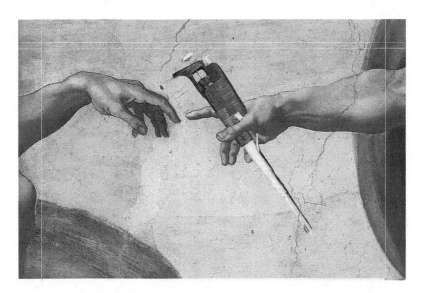

FIGURE I.2. Michelangelo's *Creation of Adam*, modified for synthetic biology. Image from Alan Moses, "Intelligent Design: Playing with the Building Blocks of Biology," *Berkeley Science Review* 5, no. I (2005). Courtesy of Tracy Powell / *Berkeley Science Review*.

were not raised in Jewish or Christian households, and many identified as
unreligious, secular, or atheist). Such ethical concerns would have been articulated and illustrated differently if they had been rooted in, for example,
Hindu or Buddhist cosmologies.[68]

In his ethnography of Artificial Life, Stefan Helmreich argues: "In their
story of how evolution has hijacked humans' working energies to engineer
the next stage of evolution, Artificial Life researchers only occasionally
notice that this narrative positions them as a new elite. When they speak
of humanity, they are speaking of a small fraction of humanity, and they
are explicitly locating themselves as the vanguard force of evolution."[69]
Joining acts of creation to partaking of forbidden knowledge—making
life to understanding life—Judeo-Christian imagery casts synthetic biologists as acting outside the "natural order," even as it also allows them to
imagine themselves simultaneously as either a "vanguard force of evolution" or explicitly invested with godlike powers (think of Endy's declara-
tion that his laboratory is the "artificial living world that [he] completely
control[s]," in which T7.1 is allowed to flourish).

Other synthetic biologists were more than willing to accept the mantle
of "intelligent designer," with all its theological connotations. Harvard
synthetic biologist George Church described his thinking on the matter,
"We're acting as engineers, possibly as intelligent designers. The religiously
inclined would not put humans in the same league with the 'Intelligent Designer,' or God. . . . We, as intelligent designers, are not in the same league
as the 'Intelligent Design' forces that started the whole shebang. . . . We're
not even designing the basic idea of life; we're just manipulating it."[70] Yet
in the same lecture, he pushed the ID metaphor further, leveraging it to
reflect on the work of synthetic biologists as a continuation of natural
selection, viewing his fellow scientists as objects and agents of natural selection: "We seem to be 'designed' by nature to be good designers. In
that sense we're part of some huge recursive design, but we're not doing
something we're not designed (and microevolved) to do. Engineering is
one of the main things that humans do well. . . . It's just what we do and it's
natural."[71]

Church's comments, which equate evolution with design and nature with
artifice, elicited this response from popular historian and essayist George
Dyson, who asked laconically, "are we learning to manipulate life or is life
learning to manipulate us?" He speculated that perhaps synthetic biologists
have been parasitized by "code-consuming and code-spewing microproces-
sors" that allow them to "help" life self-replicate into more evolved forms.[72]

Science studies scholar Richard Doyle has argued that human capacities, such as rhetoric, consciousness, and technology, sculpt biological evolution, just as humans are themselves imbricated in an ecological "involution" in which other species also exert forces, impulses, and desires that torque evolution and constitute living form.[73] Church and Dyson's staging of synthetic biologists as the unwitting vectors of biological evolution unseats synthetic biologists as creators, engineers, or designers. They suggest instead that T7 had designs on Endy and his students, rather than the other way around.

Natural, Unnatural, Supernatural

The various evolutionary tales MIT synthetic biologists tell themselves about themselves accomplish a neat rhetorical sleight of hand: synthetic biologists appear as (1) self-directed engineers who manipulate and "design" life without creating it; (2) biological beings who are fulfilling their evolutionary destiny by doing what they were "designed" by nature to do anyway, which is make "better" versions of life; and (3) "intelligent designers" with godlike powers to shape and modify life itself. Or perhaps the phage had infected them, and they were hosts in its next self-directed iteration of its own genome. Squint and tilt your head just so, and the story changes—it's all a matter of perspective.

Tracking the convergence of biological making and biological knowing as it was pursued by synthetic biologists at MIT in 2005 and 2006, especially around the T7.1 project, reveals that "better" is an ambiguous term. It refers to both biology that *functions* better (that does the things MIT synthetic biologists want it to do) and biology that is more *comprehensible* to them (systems that are easier to understand). For these synthetic biologists, a "well-designed" living thing is one that is optimized for human understanding, and most evolved organisms do not pass the biologists' comprehensibility test. The rationale for such thinking is the circularity that inheres in the convergence of knowing and making. Composition, for synthetic biologists, furthers biological understanding, but sometimes composition is also guided by understanding as *itself* a design principle.

Turning the equation life = information inside out, life in the early twenty-first century became *information's opposite*, and T7.1 was demonstrative of that fact: more noise than signal, more complexity than simplicity, more randomness than design. The computer virus is already imagined, in computer science circles, to be bad or "junk" code. The sorts of viruses made of genetic material and protein capsids are now imagined to also be bits of bad code.[74] They need to be recoded, refactored, and salvaged—to rescue life from its own proliferating misinformation. When making new biological things becomes the path by which biology is best understood, then understandable biological things get preferentially made. Exegesis becomes a selective evolutionary force.

From Still Life to More Intense Life

syn·thet·ic (sĭn-thĕt′ĭk)
adj.
(In most senses opposed to TRADITIONAL *adj.*)

2. *Art.* Compound. Of an art form (theater, dance, plastic arts), characterized by modernist approaches to joining multiple media; expressive of futurism, chaos, disorder; a rejection of tradition (Opposed to NATURAL *adj.* and ANALYTIC *adj.*)

    1915 F. T. Marinetti *The Synthetic Futurist Theatre* Our Futurist theater will be: *Synthetic.* That is, very brief. To compress into a few minutes, into a few words and gestures, innumerable situations, sensibilities, ideas, sensations, facts, and symbols.

    1920 Daniel Kahnweiler *The Rise of Cubism* Instead of an analytical description, the painter can, if he prefers, also create in this way a synthesis of the object, or in the words of Kant, "put together the various conceptions and comprehend their variety in one perception."

S ynthetic cubists arranged their still lifes by pasting the detritus of lives lived—train tickets, newspaper headlines, fabric, and wallpaper— onto their canvases. They ran combs across wet paint or stirred sand into it or doused it with varnish in order to produce illusions of fabric or wood grain. Synthetic cubism was named as such because it departed from analytic cubism, which had disassembled three-dimensional objects into abstract component shapes. Analytic cubism gave way to the synthetic as "the disintegrated image of the natural object gradually took on a more and more abstract geometrical shape," until "the geometrical shapes are so remotely related to the original form of the object that they seem almost to have been *invented rather than derived.*" Synthetic cubism shuffled and merged forms without analyzing them. It was "not a breaking down, or analysis, but a building up or synthesis."[1] This style is best represented by the work of Pablo Picasso and Georges Braque between 1913 and 1921, both of whom used techniques such as collage, *assemblage,* and *papier*

FIGURE INT-2.1. Georges Braque, *Still Life with Tenora*, 1913.

collé to decontextualize and then rearrange disparate real-world objects, painting by gathering pictorial fragments together.[2]

Why did artists like Braque graft and gum together sawdust, newspaper clippings, and metal shavings? For some artists, synthetic cubism was merely one instance of a much broader historical current moving away from description and toward composition. Cubist painter and sculptor Juan Gris, for example, wrote in the *Bulletin de la vie artistique* that the "analysis of yesterday has yielded to the synthesis of today." Comparing the transition from analytic to synthetic cubism to the advent of modern physics, he explained, "In the beginning, cubism was an analysis which was no more painting than the description of physical phenomena was physics. In its beginnings cubism was only a new mode of the representation of the world."[3] Such an analogy of cubism to physics suggests that the synthetic approach is opposed to merely descriptive inquiry, contrasting description with experimentation, the reflective judgment of natural philosophy with the empiricism of natural science.

Others, however, compared the work of synthetic cubists to the synthetic a priori. For example, Daniel Kahnweiler argued in 1920, "Instead of an analytical description, the painter can, if he prefers, also create in this way a synthesis of the object, or in the words of Kant, 'put together the various conceptions and comprehend their variety in one perception.'"[4] Synthetic, for cubists like Braque and critics like Kahnweiler (as for today's synthetic biologists), meant bringing disparate elements into proximity in order to assemble or compose—and thereby comprehend—something entirely new.

Not limited to the visual arts, the synthetic seeped also into machine-age performing arts, in particular Italian futurism and the Russian revolutionary avant-garde during World War I and the interwar period. F. T. Marinetti's 1915 manifesto "The Synthetic Futurist Theatre" "condemn[ed] the whole contemporary theater," declaiming, "Our Futurist theater will be: *Synthetic.* That is, very brief. To compress into a few minutes, into a few words and gestures, innumerable situations, sensibilities, ideas, sensations, facts, and symbols." Synthesis functioned dually as a condensation and an agglomeration of theatrical techniques aimed at achieving an art form or "deformation" approximating a "reality [that] throbs around us, bombards us *with squalls of fragments of interconnected events, mortised and tenoned together, confused, mixed up, chaotic.*"[5] It presented what Fluxus artist Dick Higgins would later describe as "performance pieces which were synthesized out of extremely raw-seeming materials."[6] As the lived world of modern urban Europeans shattered into staccato disorder and specters of war, their theater kept pace rather than clutching at nostalgia.

A similar aesthetic and politics guided Soviet "synthetic dance" in the early 1920s. Inna Chernetskaya, former student of Isadora Duncan, rejected Duncan's modernist organicism: "this notion of the flow of 'natural' dance, and indeed everything which Duncan . . . stood for, was anathema."[7] Chernetskaya founded the Studio of Synthetic Dance within Moscow's Choreological Laboratory in 1923, where she honed a choreography joining acrobatics, mechanistic movements, and mime. Comparing her studio's aesthetic to symbolic synesthesia, she explained, "I think that the path of the free dance of the future will occur in an organic fusion of the three arts of painting, music, and dance."[8] Synthetic, for modernist dramatists, performers, and choreographers, meant rejecting an unproblematic naturalism or organicism in favor of a postnatural concatenation of diverse images, forms, techniques, and media.

This valence of synthetic is also apparent in the work of synthetic biologists, who build biosynthetic metabolic pathways and living systems that

incorporate genetic material from diverse phyla and species. Synthetic biologists are magpies of the living world, pasting together "found" protein-
coding regions from diverse organisms—wormwood trees, bacteria, and
yeast, for example—in order to make something new. And like synthetic
cubism, theater, and dance, the resulting artifacts undermine distinctions
between reality, organicism, and representation, as when *papier collé*
paintings reject mimesis in favor of either presenting abstract signs for the
real or turning real objects into representations simply by slapping them
onto canvas. By what strange logics might we nominate and classify the
resulting compositions? Following on the still lifes of a century ago, one
signpost cubists might offer biologists is that mixing media is a technique
by which "to discover the true relationships among objects, subsuming
several qualities into one higher reality, but [with] the higher reality . . .
still assumed to lie behind the world of objects."[9]

CHAPTER TWO

The Synthetic Kingdom: Transgenic Kinship in the Postgenomic Era

The genes and supergenes indulge occasionally in queer freaks and lapses. —Jacques Loeb, *The Organism as a Whole*, 319

My genes are point zero three percent *Cyprinus carpio*—freshwater carp. I'm a patented new fucking life form. —Larissa Lai, *Salt Fish Girl*, 158

From Taxonomy to Technique

"The Synthetic Kingdom: A Natural History of the Synthetic Future" (fig. 2.1) diagrams biological kingdoms forking and spiraling from a central "root." Explicitly drawing on microbiologist Carl Woese's 1990 tripartite division of life on earth into the domains Bacteria, Archaea, and Eucarya, "The Synthetic Kingdom" illustration adds a fourth branch to Woese's model, a branch designated Synthetica. Each color-coded kingdom in turn branches to map common ancestry of related phyla. Shorter branches identify more evolutionarily recent divergences. Eucarya, for instance, is a whorl encompassing genera as diverse as humans and paramecia. Woese premised his 1990 phylogenetic model of microbial biological relatedness on 16S ribosomal RNA (rRNA) sequences, subunits in prokaryotic ribosomes whose cellular function is primarily structural.[1]

This fourfold diagram does not, however, cleanly follow molecular genetics, at least not when it comes to its fourth domain, Synthetica. For this tree of life is the work not of a molecular biologist but of artist Daisy

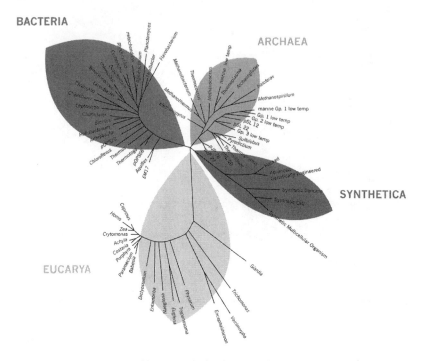

BACTERIA

ARCHAEA

SYNTHETICA

EUCARYA

FIGURE 2.1. Alexandra Daisy Ginsberg, "The Synthetic Kingdom," 2009. Courtesy of the artist.

Ginsberg. The fourth kingdom, which she added to the scientifically recognized three identified by Woese, currently includes five phyla, all lacking Latin nomenclature. Ginsberg identifies these phyla simply as "hacked," "advanced genetically engineered," "synthetic genome," "synthetic cell," and "synthetic multicellular organism." "The Synthetic Kingdom" is part of a body of work in which Ginsberg combines synthetic biology with artistic practice.[2] Yet "The Synthetic Kingdom" is not only an artistic provocation. Synthetic biologists have debated Ginsberg's taxonomy, which appeared as the cover image of a 2010 special issue of *Nucleic Acids Research* and of *Synthetic Biology: A Primer*, a textbook for undergraduates and incoming graduate students studying bioengineering.[3]

Ginsberg narrates her diagram in a phylogenetic idiom. She identifies it as the latest in a long line of trees of life, among whose devisers she counts key figures in the modern history of taxonomy: Carl Linnaeus, who differentiated Plantae and Animalia in 1735;[4] Ernst Haeckel, who in 1866 added Protista to the mix;[5] Edouard Chatton, who split the living world between prokaryotes and eukaryotes in 1938;[6] Herbert Copeland, who in

1956 identified four biological kingdoms;[7] Robert Whittaker, who counted
fungi among five kingdoms in 1969;[8] and Woese's tripartite model in 1990.[9]
Yet the living world is no longer what it was in the eras of Haeckel or Copeland, and new life-forms could potentially throw a spanner in the phylogenetic works. In her artist's statement, Ginsberg asks: "How will we classify
what is natural or unnatural when life is built from scratch? . . . Biological
taxonomy is, after all, an artificial construct. We will have to insert an extra
branch into the Tree of Life. The Synthetic Kingdom is part of our new nature. How else will we make sense of this novel menagerie of engineered
life?"[10]

   Taxonomy tames novelty by incorporating it into a predetermined and
codified system. However, the rules governing *how* and *why* living things
are classified in certain ways are changeable.[11] The ancients, from Pliny
and Galen to Aristotle, believed the natural world to be static and tending toward perfection. Medieval natural philosophers surveyed in wonderment *naturalia* and *artificialia* gathered into cacophonous curiosity cabi-
nets and private collections, musing over monsters and other marvels.[12]
European travelogues and bestiaries were exhaustively detailed but sel-
dom orderly, with entries ranging from an organism's morphology to recipes incorporating it. From the age of exploration through the Victorian
era, imperial, scientific, and commercial voyages returned to European
ports groaning with fantastic new zoological and botanical specimens that
boggled imaginations and tested the limits and flexibility of taxonomic
systems.[13] Carl Linnaeus, whose name became synonymous with modern
systematics, proposed a self-consciously artificial order that used binomial
nomenclature in hopes that it would be handier and more memorable
than prior systems.[14] It was not until the twentieth century that taxonomy
transformed from natural history to experimentalism, as biologists used
the tools of molecular evolution and bioinformatics to compare proteins
and nucleic acids.[15] Only in the most recent decades have biological engineers and biotechnologists been tasked with classifying organisms that
they have purposefully engineered to amalgamate genetic material found
in distant branches of the phylogenetic tree.

   To ask an old anthropological question of a new biotechnical phenomenon: by what genealogical logics do synthetic biologists classify and sort
engineered artifacts that combine genetic parts from diverse species? More
broadly, what models of taxonomy and phylogeny are currently placed
under pressure by the manufacture of new synthetic biological systems?
Harking back to Michel Foucault's claim that biology is a discipline founded

upon the choice to divide up the world into the living and the nonliving, we may ask how the discipline of biology changes when bioengineers actively query the natural and the unnatural. Earlier biological classifications have been premised on external physical appearance, patterns of cellular organization, common ancestry, behavior, and microbial phylogenetics, but some synthetic biologists are dynamically revising such classifications. For them, *technique becomes taxonomy.*[16]

Transgenic critters, such as strawberries bearing fish genes, have been troubling categories of relatedness—species, lineage, and consanguinity—since the mid-1970s, as scholars such as Sarah Franklin and Donna Haraway have noted.[17] They "fit into well-established taxonomic and evolutionary discourses and also blast widely understood senses of natural limit. What was distant and unrelated becomes intimate."[18] This trend is especially apparent in synthetic biology because its practitioners take transgenic technologies to radical extremes of genetic hybridity, and because they oscillate between characterizing the resulting organisms as "natural," "unnatural," and "postnatural." Indeed, some synthetic biologists question "species" as a coherent category altogether, suggesting instead that their work ruptures species, both molecularly and discursively.

Jacques Loeb in 1916 had something similar in mind when he noted the astonishing fact that some organisms display galvanotropism (movement toward an electric current), even though organisms are likely to be exposed to a constant electrical current only in a laboratory. To explain this seemingly purposeless capacity, Loeb throws up his hands, admitting that genes "indulge occasionally in queer freaks and lapses."[19] While Loeb here invoked queerness to mark the surprising consequences of placing "natural" organisms in "unnatural" scenarios, the term is theoretically generative in other registers.

Natural: Classifying Catalyzing Chimeras

The Bay Area is home to synthetic biologists trained in chemical and metabolic engineering, who, more often than not, compare cells to factories and call themselves "the blue-collar workers" of synthetic biology. In synthetic biology labs ranging from academic laboratories to large companies to small start-ups, as well as labs funded by the Defense Advanced Research Projects Agency (DARPA), synthetic biologists insert between six and twenty genes from extant species into either *E. coli* or yeast, working to turn cells into microbial "factories" that produce drugs and fuels. The Joint Bioenergy Institute (JBEI) is a spacious, glass-walled complex in Emeryville located in the same building as Amyris Biotechnologies. JBEI's well-manicured main courtyard is punctuated by modernist sculptures and murmuring fountains. To get there from my hotel, I would walk along an overpass above train tracks snaking toward San Francisco, passing a Chevron station where the cost of gas that week was $4.27 a gallon. The station— and the price of gasoline—are visible from all of the building's south-facing windows, a reminder to researchers of the price point they need to meet in order to make biofuels economically viable. JBEI researchers are building thousands of strains of microbes, each of which contains nine genes, give or take a few, from plants, yeast, and bacteria, in the service of digesting plant cellulose (like corn or switch grass) to produce fuel precursors. When President Obama promised college students in his 2011 State of the Union address that they will get jobs in the biofuels sector in a few years, calling biofuels one of the "Apollo Projects of our time," JBEI is exactly the kind of thing he was talking about.

These synthetic biologists frame their hybrid, engineered organisms as nothing that nature does not already do. Owen grew up in a small town in New England, the first in his family to attend college. He has worked at JBEI for two years after earning a PhD in the life sciences at an Ivy League university.[20] There, his thesis focused on building microorganisms that break down fuels and other pollutants. The irony of his career, he tells me, is that now he uses microbes to build new fuels instead of break them down. Sitting beneath the incandescent glare of a meeting room at JBEI, he diagrams for me a bacterial metabolic pathway for producing fuel precursors. "And this is where the magic happens," he says excitedly. "We take these genes from plants, from yeast, and native *coli* genes and we assemble them in a plasmid" (fig. 2.2).

Part of Owen's job is to get genes from diverse organisms to work in concert so that he can maximize fuel production per cell. Owen puts in long hours at the lab, staying up for seventy-two hours at a time to sample his engineered bacteria's chemical output every other hour, napping fitfully in one of the lab's warm rooms. He tells me that if he stops to think about what he does all day, he finds it "bizarre," building a "chimera, all these organisms together." And when he talks to his parents about his job, he realizes how unnatural it sounds to the uninitiated—his mom worries that the bacteria her son builds might accidentally be released into the environment and wreak havoc. Still, he admits, he rarely reflects on the unnaturalness of

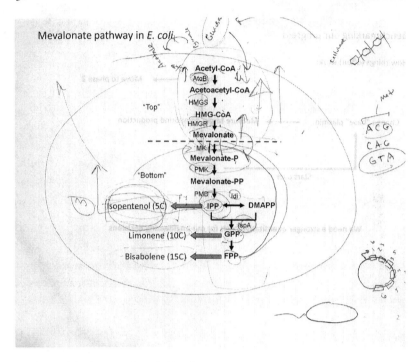

FIGURE 2.2. Mevalonate pathway in *E. coli.*

his organisms. Most of the time he does not think of his labor as fundamentally different from anything farmers and breeders have done for millennia, or from what biology does on its own by means of lateral gene transfer (the phenomenon by which genetic material moves horizontally between organisms via viral vectors, cell-to-cell contact, or the uptake of naked genetic material). In his view, he merely "facilitates" a process that would otherwise take nature billions of years to accomplish. He explains, "it's just DNA. It's the same As and Cs and Gs and Ts as this [other] piece of DNA," and "at the molecular level, they're just the same, right?" His question feels both rhetorical and plaintive—he wants me to reassure him that all that DNA really is "just the same," that nucleotides have themselves become the universal substance rooting all earthly relatedness, thereby erasing any meaningful difference between species, class, or clade, as well as between "natural" and "unnatural" (i.e., human-directed) genetic revisions.

I heard similar language from dozens of other synthetic biologists. The same week, while eating paninis for lunch in a cafeteria on the ground floor of an airy atrium in a nearby office building, Peter Ackermann, an industry liaison at the Synthetic Biology Engineering Research Center (SynBERC), compared synthetic biology to the Gardens of Versailles, telling me,

> The thing is, people have been changing nature since the dawn of civilization. The first farmers and herders were also our first genetic engineers. Think of the Bible—it describes how Jacob genetically engineers goats! Modern bioengineering could be compared to the Gardens of Versailles. In their time, Versailles made a dramatic impression on visitors. The king of France had taken a swamp and wild forest and turned them into a completely artificial, designed garden. Perfectly cut hedges and geometric lakes. He was telling the world that he, Louis XIV, was not just the absolute ruler of people; he ruled nature. And these days a visitor might think, well, it's a garden for Chrissake.[21]

Ackermann concluded, in language I often heard invoked by other synthetic biologists, "Mutations happen all the time in *nature*—randomly, sexually, even via horizontal gene transfer between very different species. You might know the latter as 'transgenics.' People have actively promoted such mutations and deliberately selected beneficial traits for millennia. Modern bioengineering continues this tradition with much greater precision and understanding of what it is we are doing."

The previous summer at a synthetic biology conference in London, an executive of a large synthetic biology corporation offered me a more nuanced account of how synthetic biologists split the difference between nature and artifice. He told me that synthetic biology is "schizophrenic": "when we're talking to funders, we say synthetic biology is fundamentally different from biotech because we're making new things never found before in nature. But when we're talking to others, like the press, we say everything we're doing is *natural* and boring. We're just using natural things like plants, sugar, and yeast to make something useful."[22]

Synthetic biologists building cascading synthetic chemical reactions, or "biocatalytic pathways," both at JBEI and at nearby labs, place their work on a continuum with natural selection. They claim that by shuffling genes between diverse species, taxa, and kingdoms, they are only doing more of what nature already does when genes move laterally between species.

Yet in another register, they think about what they are doing as, to quote one synthetic biologist, "never found before in nature." This wavering between novelty and conventionality, especially as I listened to synthetic biologists articulate it in the summer of 2013, put me in mind of something seemingly unrelated: American LGBTQ politics, which had triumphed on multiple civil rights fronts by leveraging rights of access to highly conventional, even traditional or conformist, forms of life (monogamy, military service, taxation, citizenship, etc.). Though in 2009 synthetic biologists intentionally referenced political and legal debates over intelligent design, in 2013 they did not knowingly draw upon queer politics to describe the organisms they designed. Rather, it is a discourse in which I am complicit. At the time, I noted that in both domains (LGBTQ rights movements and synthetic biology), it is sometimes politically useful to neutralize new forms of life by enrolling and impaneling them into social forms already accepted and glossed as unproblematically "acceptable" or "normal" (read: "natural").[23]

Denaturalized: Affinity Undone

Cultural anthropologists have examined how cultural practices are consolidated and warranted by their framing as normatively or hegemonically "natural." In particular, they have analyzed kinship as a cultural system that merges blood (consanguinity) with marriage (affinity), nature with culture. Kinship studies were revived by anthropologists in the late 1960s and 1970s, most notably by David Schneider and Gayle Rubin.[24] Until Schneider's intervention, anthropologists had treated biogenetic relatedness— genealogy—as the inviolable and universal ground by which kinship is defined. Relatedness may be socially organized and sanctioned, but such cultural prescriptions and proscriptions are always, in this view, premised in blood. Schneider upended such thinking by approaching kinship as a symbolic system like any other. He denaturalized kinship by problematizing how relatedness is defined according to rules of heredity, law, and custom. For Schneider, even adoption, an unassailably legal form of kinship, is patterned in the United States on biogenetics. Biological parentage is the metaphor that gives cultural meaning and legal weight to adoption. In demonstrating this to be the case, he showed that cultural anthropologists are also prone to thinking inside folk theories, even grounding American kinship on biology and genetics and reframing it in response to new developments in the life sciences.

Seven years later, Rubin revisited the work of Claude Lévi-Strauss, Karl Marx, and Sigmund Freud, intercalating the three thinkers to synthesize an unorthodox reading that denaturalized the sex/gender system.

She demonstrated that kinship relations are economic transactions in which a heterosexual woman functions as a "conduit of a relationship rather than a partner to it."[25] By identifying women as goods, Rubin evaluated kinship as a form of reciprocal exchange that, by extension, naturalizes heterosexuality.[26] She suggested in closing that feminist anthropologists reappraise how gender and labor relations are systems that support one another.

How, then, might one account for the ways in which synthetic biologists and allied researchers puzzle through transspecies relatedness when bits of genetic material from multiple organisms commingle in the products of synthetic biology? Kinship theory, which probes the ways people parse nature and culture and understand lineage, inheritance, blood, and property, has much to offer scholars of contemporary life sciences.[27] In turn, the novel life-forms emerging from the labs of synthetic biologists require that anthropologists revise kinship theory in surprising ways.

Unnatural: Common Ancestors and Promiscuous Families

If Bay Area synthetic biologists are "blue-collar workers," Jay Keasling is the management. Keasling, who runs both JBEI and Amyris Biotechnologies, is unlike the MIT crop of synthetic biologists. For one thing, the US Patent and Trademark Office lists him as an inventor on twenty-three patents; filing patents on the products of synthetic biology was anathema to most members of MIT's Synthetic Biology Working Group. Keasling grew up on a corn and soybean farm in Harvard, Nebraska, which has been owned by his family for four generations. In an interview with Neil deGrasse Tyson, Keasling recalls spending the first eighteen years of his life with "the smell of pig manure on [his] hands."[28] When a reporter asked him why he did not stay on the farm to become the fifth generation to steward his family's land, he joked that the stork must have dropped him at the wrong address.[29] Burdened by growing up queer in a small midwestern town, Keasling remained closeted throughout high school, college, and graduate school, only coming out to his father when he found the right address, as a professor at the University of California, Berkeley.

He has traveled far from Nebraska, personally and professionally, yet crops like corn and sugarcane remain central to his research, which in recent years has gained him notoriety. In 2006 *Discover* magazine named him "Scientist of the Year," and he, like fellow synthetic biologist George Church, has been interviewed by Stephen Colbert on Comedy Central's

Colbert Report. In the interview, Colbert, playing his neocon alter ego, accused Keasling of "playing God," until Keasling told him that he was only
modifying a process nature already accomplishes: microbes metabolize
sugars to produce ethanol. Colbert countered by berating Keasling for
allowing God to "beat him to the punch."

Now a professor in the Department of Chemical and Biomolecular Engineering and the Department of Bioengineering at Berkeley, Keasling
has served since 2009 as the CEO of the Department of Energy–funded
JBEI.[30] He is also the director of both the Physical Biosciences Division of
the Lawrence Berkeley National Laboratory and SynBERC, the DARPA-
funded synthetic biology research initiative. But the fanfare—his turn on
the *Colbert Report*, his award from *Discover*—was the result of one project
that attracted media attention in the mid-2000s. Keasling led researchers
at Amyris in manufacturing a pathway—what chemical engineers term
a "synthetic metabolic pathway for biocatalysis"—that inserted eight ad-
ditional genes from multiple organisms into yeast cells in order to produce
artemisinic acid, the precursor to artemisinin.

Between 225 and 400 tons of artemisinin are required annually to treat
malaria, and until the synthetic pathway was constructed, all artemisinin
had to be isolated from sweet wormwood (*Artemisia annua*), a plant native
to Eurasia. Chinese medical practitioners have long used this plant, called
qinghao, to treat fevers. Indeed, it was adopted as part of the Western
pharmacopoeia only in the 1960s and 1970s, as Chinese scientists, working
alongside traditional medical practitioners in the Academy of Traditional
Chinese Medicine under Maoist control, began sifting through materia med-
ica of traditional Chinese medicine in search of new pharmacologically active disease treatments. They identified the active molecule in *qinghao* in 1971,
and pharmaceutical companies began marketing artemisinin in its isolated
and purified form in the 1990s. By 2004 the drug was widely promoted by
the World Health Organization as an effective treatment for malaria in
equatorial regions.[31]

Malaria to this day is one of the greatest threats to global health, dis-
proportionately affecting populations in tropical climates. The disease is
contracted via mosquito bites, as mosquitoes infected by *Plasmodium falciparum* transmit the parasite to humans. After decades of treatment, *Plasmodium* is often resistant to multiple drugs, including chloroquine and
sulfadoxine-pyrimethanine. Artemisinin, to which *Plasmodium* has not yet
developed resistance, remains one of the most effective drugs in the World
Health Organization arsenal, used either alone or in combination with

other therapies. Keasling set out to synthesize artemisinin synthetically, a feat that would effectively bypass the production limits and uncertainties of harvesting sweet wormwood, the supply of which is dependent on climate fluctuations, political conflict, and trade embargoes.

Keasling's project began in 2000 when one of his students showed him a journal article about a molecule called amorphadiene. At the time, Keasling was casting about for an interesting molecule on which to test his theory that he could cobble together synthetic pathways to compel microbes to synthesize a potentially endless number of naturally occurring compounds. Once he had chosen artemisinin as his target, the lab needed to be able to identify enzymes involved in the metabolic pathway that produces artemisinin. By 2003 Keasling and other researchers at Berkeley and the Lawrence Berkeley National Laboratory reported that they had inserted a pathway from yeast and multiple genes found in other organisms into *E. coli*. Bolstered by these initial results, in December 2004 Keasling and his team were awarded a $42.6 million grant from the Bill and Melinda Gates Foundation.

With this initial infusion of cash, Keasling founded Amyris Biotechnologies, a not-for-profit company that aimed to engineer microbes that could manufacture artemisinin more cheaply than naturally occurring artemisinin. The foundation gave them three years to complete the project. Researchers at Amyris began by assuming that they could seek such enzymes not only in wormwood but also in plants related to wormwood. As they described it in an idiom inflected by the language of relatedness, "We hypothesized that plants belonging to the Asteraceae family would share common ancestor enzymes," and that such enzymes would be "highly conserved in three distantly related genera in the Asteraceae family, but not in plants outside the Asteraceae family."[32] Seeking catalytic enzymes in related species, they checked sunflower and lettuce plants for the presence of genetic expressed sequence tags (ESTs) for a particular catalytic molecule.

Scientists in Keasling's lab at Berkeley collaborated with researchers at Amyris Biotechnologies. They began working with *Saccharomyces cerevisiae* (brewer's yeast), inserting DNA coding for amorphadiene synthase, an enzyme that catalyzes the molecular pathway for artemisinin synthesis. They then modified the transgenic genome of yeast bearing wormwood genes to increase 500-fold the production of a molecular precursor to artemisinin, artemisinic acid. The new strain of yeast yielded as much artemisinic acid relative to biomass as wormwood but at a much greater rate than wormwood. While wormwood takes several months to grow and produce artemisinic acid, the synthetic yeast accomplished the same feat in a few days.

In his study of a malaria outbreak in Egypt in the 1940s, historian Timothy Mitchell weaves together a drama enacted by *Anopheles gambiae* mosquitoes, *Plasmodium falciparum*, and Egyptians, demonstrating that "[d]ams, blood-borne parasites, synthetic chemicals, mechanized war, and man-made famine coincided and interacted," and that "the linkages among them were hydraulic, chemical, military, political, etiological, and mechanical."[33] The story I tell here also follows jungly pathways of interspecies interaction as bacteria, yeast, and sweet wormwood are bound together in the Berkeley laboratory of a former Nebraska farmhand, with the help of historical contingencies ranging from Maoist cultural agendas to the commercial success of Microsoft, to yield a drug that is able to prevent epidemics that flourish at the confluence of *Anopheles* and local war, *Plasmodium* and polluted water. The "unnatural" biocatalytic metabolic pathways sutured together by chemical engineers-turned-synthetic-biologists mirror the "messy world of reality," the "hybrid agencies, connections, interactions" that constitute human history.[34] New forms of genetic relatedness may be engineered in synthetic biology laboratories, but they underscore already-complicated relations between the natural and the artificial at work in the lived world.

Amyris freely licensed this transgenic synthetic pathway to Sanofi-Aventis, a French pharmaceutical company, for mass production and distribution in developing countries. After approval by the World Health Organization in May 2013, Sanofi teamed with Huvepharma, a Bulgarian pharmaceutical factory, which produced 28.7 tons of artemisinic acid in 2012 and predicted a 2013 target of 60 tons (yielding 35 tons of artemisinin after it is purified by a light reactor in Garessio, Italy). Sanofi brought synthetic artemisinin to market in August 2014, when it shipped an initial distribution of 1.7 million doses to malaria-endemic countries in Africa. With the press and attention garnered by this trick, which lowered the cost of artemisinin tenfold, from US$2.40 to approximately 25 cents per dose,[35] Keasling moved on to his next big biosynthetic project: biofuels.[36]

The ways in which bioengineering messes with neat distinctions between the natural and the unnatural are not lost on Kristala Prather, one of Keasling's first graduate students at Berkeley. Now a tenured associate professor of chemical engineering at MIT and a rising star in synthetic biology, her awards and distinctions include a fellowship at the Radcliffe

Institute and a turn as a speaker at the World Economic Forum. She spent four years working at Merck after completing her PhD and before joining MIT's faculty in 2004. I first met her in 2008 on a plane on our way to a synthetic biology conference in Hong Kong, where she was accompanied by her husband and two small children. To our mutual delight, we discovered that we had grown up not far from each other in east Texas, tipped off by fast-fading accents.

The morning after our flight to Hong Kong, while I was still bleary-eyed from an eighteen-hour flight and a few hours of shallow sleep, Prather was not only awake but also magnetically enthusiastic. She, like her former adviser, develops biocatalysis pathways to manufacture what she terms an "unnatural product spectrum." Simply put, she says, she wants to combine chemical engineering with synthetic biology, getting cells to make chemicals previously found only in other organisms (such as plants) or unknown in nature. On the first morning panel of the three-day conference, Prather opens her talk before an overflowing lecture hall by describing her efforts to synthesize a compound called D-glucaric acid, a chemical that has no known naturally occurring microbial pathway. She frames her lab's work to synthetically hybridize a pathway combining enzymes found in yeast, mice, and the bacterium *Pseudomonas syringae* as a bad joke about novel transgenic kinships: "A yeast cell, a mammal, and a bacterium walk into a bar." She waits a beat, pausing for audience reaction. "And out comes D-glucaric acid." She laughs out loud at her own joke, despite being met with groans from her audience.

Unlike other synthetic biologists at MIT, and perhaps because of her training in chemical engineering and industry, Prather does not see her work as being an extension of computer and electrical engineering principles into the natural world. In July 2012, over lunch at a Kendall Square pub in Cambridge known for its extensive beer list, she tells me that she considers her work to be much closer to the historical origins of industrial chemical engineering in the late nineteenth and early to mid-twentieth centuries, in which microbes were coaxed into manufacturing beer, explosives, and drugs like penicillin. The language in which she describes the construction of such synthetic metabolic systems relies heavily on descriptions of kin and kind, biological associations and unnatural pairings. She uses the terms "unnatural," "synthetic," and "nonnatural" interchangeably when referring to the construction of newly combined cellular pathways for the synthesis of compounds.

Such language is even more evident in her published papers, in which

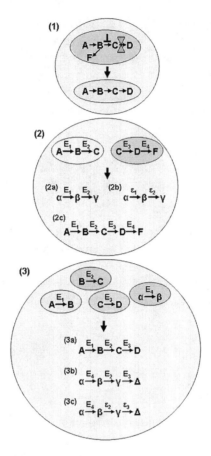

FIGURE 2.3. "Strategies for Synthetic Pathway Creation Arranged in Increasing Degrees of Departure from Nature." From Collin H. Martin, David R. Nielsen, Kevin V. Solomon, and Kristala L. Jones Prather, "Synthetic Metabolism: Engineering Biology at the Protein and Pathway Scales," *Chemistry and Biology* 16, no. 3 (2009): fig. 4.

pathways already inherent in a cell are "native" and "produce natural bio-products." However, "in cases where natural evolution has fallen short of industrial needs, the tools and practices of synthetic biology can be applied to aid in the creation of designer enzymes and cellular phenotypes."[37] In a review paper Prather coauthored, she delineates multiple experimental approaches used by synthetic biologists working in metabolic engineering; in one figure she sorts such strategies according to "increasing degrees of departure from nature" (fig. 2.3). The most extreme of these approaches, which both she and Keasling use, is the construction of what she terms "truly unnatural" metabolic pathways "from multiple, ordinarily unrelated enzymes."[38] Listing examples of representative work done by her colleagues, she calls upon a diverse menagerie of organisms that are stitched together into new living systems. Latin names proliferate alongside chemical nomenclature as chemical engineers and synthetic biologists harvest genetic material from *Corynebacterium glutamicum, Bifidobacterium lactis, Pseudomonas fragi, Pseudomonas putida, Amycolatopsis orientalis, Streptomyces coelicolor*, and *Porphyromonas gingivalis*.

Such pathway construction relies on what molecular biologists term "enzymatic promiscuity,"[39] another word implying indiscriminate and casual couplings and admixture. Promiscuous enzymes are a boon to metabolic engineering, as they can catalyze multiple reactions and show preliminary evidence of "speeding up" evolutionary change. In a *Nature Chemical Biology* paper, one biochemist classed enzyme promiscuity as an example of "messy biology," explaining: "Biological messiness relates to *infidelity, heterogeneity*, stochastic noise and variation."[40] Similarly, recall Keasling's work to select those "ancestor enzymes" common to the Asteraceae family, choosing those enzymes that are "conserved in three distantly related genera in the Asteraceae family" and that therefore share a close lineage with similar enzymes.

"Unnatural" blurs the categorical and the normative—it refers simultaneously to that which is counter to nature and to that which is against the (moral) natural order, something strange or out of the ordinary. In this regard, it bumps up against a similarly semantically charged word: "queer." The "unnatural" biocatalytic pathways that synthetic biologists and other biotechnologists cobble together[41] have much in common with queer kinship categories as analyzed and theorized in recent years by anthropologists. Queer kinship literature arises from studies of what was once termed "fictive kinship." "Fictive," of course, suggests something is artificial, synthetic, or pseudo-. In so doing, the term posits that there is a more "real" or authentic form of kinship that "fictive" kinship can only ape.[42] Yet synthetic biologists building metabolic pathways actively unsettle the relation of the biological to the "natural," whether with a capital or a lowercase *N*. That is, they *denaturalize* biology in the same way that cultural anthropologists have toiled to denaturalize categories like gender, race, and kinship.[43]

So too, new biotechnologies, like queer kinship, have forced feminist science studies scholars to appraise categories previously taken for granted—foremost among them being biology and "nature"—as unproblematic grounds upon which to define relatedness.[44] Sarah Franklin evaluates how, as "biology has become more visibly and globally dominant as a science in the second half of the twentieth century—a transformation that has seen the biology of plants, animals, humans, and microorganisms become more technologically mediated and amenable to reconstruction"—the effect has been that "the domain of the biological is today more visibly 'constructed' than ever before. This has consequences not only for how we think about biology, biotechnology, and our relations to them but also for how we figure what counts as a biological tie."[45] Appraising how new reproductive technologies like in vitro fertilization and gestational surrogacy can trouble Euro-American kinship categories, she further argues that "the ways in which humans are today connected and related through biology *undoes the very fixity that the biological tie used to represent.*"[46]

With their manufactured cells, synthetic biologists (and chemical engineers) like Keasling and Prather undo regnant models of taxonomy that link biological organisms according to lineage, descent, blood, or gene. They instead couple diverse genetic parts, mixing together that which is genetically unrelated and entangling nucleic acids in ways they alternately label "unnatural" or "nonnatural." Like queer kinship, the biotechnically enabled cells housing microbial pathways that make artemisinin, jet fuel, or other "value-added" chemical compounds are marked as transgressive, nonreproductive, and unnatural. And like queer kinship, such organisms operate outside the bounds of nuclear, heterosexual familial relations: "chosen families are neither imitative nor derivative of the dominant model of American kinship. Rather . . . they constitute a distinctive form of kinship, constitutive rather than analogous to straight kinship."[47] These organisms are often modeled on heterosexual forms of relating: they are described by their makers as "sharing a close lineage," being part of the same "family," or being descended from the same "ancestor," even as they nonetheless participate in "promiscuous activities."

In short, filtering synthetic biologists' understandings of synthetic relatedness through the lens of kinship theory reveals three effects of such thinking. First, "natural" and "unnatural" are categorically insufficient terms for synthetic biologists to describe the organisms they manufacture, even as they deploy those words to serve pragmatic and political functions. Second, definitions of "species" as a biogenetic category are put under pressure when synthetic biologists sort transgenic species, which, for them, give the lie to "species" as a coherent category in the first place. Finally, biological relatedness gets defined, not by genes, but by a "postnatural" constellation of criteria, including social exchange, rational design, and mutual obligation.

Postnatural: A Transgeneography

When I interview him in his office in the MIT Synthetic Biology Center in the summer of 2012, synthetic biologist Chris Voigt, a soft-spoken and earnest engineer, skirts the issue of the natural altogether. Voigt, at the time an associate professor at MIT, is perhaps best known beyond the synthetic biology community for the "E. coliroid" project, a proof-of-concept bacterial edge-detection system that is so named because the bacteria change color in response to a pattern of light exposure, generating a bacterial "lawn" that can replicate a black-and-white image. He describes his work and that of his colleagues, not as "natural" or "unnatural," but as "postnatural."

He shows me a different sort of phylogenetic tree: the logo of the Center for PostNatural History, a Pittsburgh museum dedicated to transgenic species. The museum was founded and is curated by bioartist Rich Pell, a high school friend of Voigt's. Pell was inspired to open the museum, Voigt tells me, after the two renewed their friendship at their ten-year high school reunion. Voigt then invited Pell to learn more about bioengineering by tagging along with him to the International Genetically Engineered Machine (iGEM) competition, an annual synthetic biology event.[48] The Center for PostNatural History logo shows the tree of life, an arrow arcing between two separate branches as a visual mnemonic for genetic transplants. It also resembles a midsagittal slice of a human brain, suggesting that human mental capacity and ratiocination are the driving forces making possible such genetic splicings. Pell describes the logo as a visual rendering of "transgeneography."

Other synthetic biologists speak even more directly about transgeneographic muddlings.[49] An East Coast–trained synthetic biologist who builds bacteria that generate electricity for Lawrence Berkeley National Laboratory put the issue in terms of the relation of dogs to wolves. In her mind, little more separated electrical E. coli from their "native" counterparts than separated dogs from wolves. She explained to me in a turn of phrase echoing Owen's earlier statement, "Yeah, but it's all As and Ts and Cs and Gs and what's the difference between a dog and a wolf? It's not that much. It's true we're taking genes from one organism to the next. But

FIGURE 2.4. Logo of the Center for PostNatural History, Pittsburgh, Pennsylvania. Courtesy of Richard Pell.

I guess part of it's also an understanding that all organisms are fundamentally all related. So you're just going back in time or you're going forward. You're taking genes from either an ancestor or a descendant and putting [them] back in. The number of cases where we're making completely artificial proteins, things nature's never made, is not that large."

While this reading of synthetic biology as "postnatural" begins with the premise that bioengineers can import genes from one species into another, rendering an organism that is an amalgam of multiple species, other synthetic biologists suggest that DNA synthesis and unnatural biocatalytic pathways rupture "species" altogether. In an age in which synthetic biologists port synthesized genetic segments from one organism into another, lineage and descent cease to be sensible, as difference dissolves into unmarked, undifferentiated chains of nucleotides.

This stance was most clearly articulated to me in August 2012 by synthetic biologist Adam Arkin. I meet Arkin at KBase (KnowledgeBase), a Department of Energy–funded project to build a unified open-source archive of genomic and bioinformatic data for systems biology. KBase is still ramping up in an office complex across the street from the Emeryville building that houses the JBEI and Amyris labs. I find that it has many more computer terminals than employees to occupy them. The few I see are scattered around an open room, typing away at their desktops while eating leftover pizza and sandwiches Arkin has ordered for them. Although a professor of bioengineering at the University of California, Berkeley, director of the university's Synthetic Biology Institute, and a director at the Lawrence Berkeley National Laboratory, Arkin is sartorially still in his postdoc years. He cuts an imposing figure, tall, stout, and bearded, dressed in jeans and a faded brown hoodie. He speaks manically, almost frenetically—a colleague told me he sounds like a cross between James Woods and John Turturro. I spend the first half hour of our conversation standing, as Arkin is too wound up to remember to sit down.

While talking about using RNA to install new functions in bacteria and viruses, we swerve into a molecular riff on Mary Douglas, the midcentury structural anthropologist who argued that religious dietary restrictions (as outlined in Leviticus) are based on beliefs about species hybridity and ambiguity: "those species are unclean which are imperfect members of their class, or whose class itself confounds the general scheme of the world."[50]

Arkin, who is Jewish, recalls for me an argument he had with a representative from a nongovernmental organization concerned with the societal impact of genetically modified foods: he was asked whether a tomato with a pig gene was kosher. While this could be a bit of Talmudic quibbling over Levitical distinctions between clean and unclean, creeping and flying animals, he uses this example to reflect on his own work in synthetic biology: "There's a reductio ad absurdum there. What makes a pig gene a pig gene? I could find a pig gene that's 98 percent similar to a human gene and 99 percent identical to a bird gene. So why is it a pig gene? What makes it a pig? And they'll say, well, it has to have pigness. Well, what is pigness? They can't answer that because it's not a meaningful question. Does a gene have pigness? No." Pursuing this line of reasoning further, Arkin asks whether a synthetic pig gene is a pig gene, concluding: "It's not the same material. It didn't come from a pig—no carbon from the original pig. How about half a pig gene? What about a base pair? When does *it* become *this thing*?"[51] DNA synthesis is the foundational technology of synthetic biology, allowing researchers to specify a genetic sequence and have it physically synthesized, nucleotide by nucleotide, for a few cents per base. Divorcing genetic sequence from a host organism, for Arkin, is the technique that scrambles species boundaries.

Answering his own question, Arkin contends that *nothing has pigness* or any other sort of species specificity, because genes can now be spit out of a DNA synthesizer rather than being sourced in a whole organism. By this logic, there is nothing particularly hybrid about a bacterium bearing genes culled from yeast, petunias, or Icelandic hyperthermophiles, because it is all just so many nucleotides. For synthetic biologists, if synthetic metabolic systems combine genetic material from dozens of phyla, it is not because they are more artificial or concatenated than other living things. Instead, they merely give the lie to the idea that *any* organism is inviolable or natural. Indeed, synthetic biologists working on such systems argue that their products are on a spectrum with the rest of the living world—there is, they point out, no organism that is not already transgenic. The transgenic products of synthetic biology call into question the notion that naturally occurring organisms are not similarly already cobbled together.

Synthetica Redux: Beyond Nature

I first conjured Ginsberg's tree of life as a discursive specimen, one that concretizes and condenses structures of thinking, feeling, and acting disclosed by synthetic biologists. I reinvoke it to forward my claim that synthetic biological entities inaugurate new modes of relatedness.

On his blog, Camille Delebecque, a Harvard-trained synthetic biologist who launched a synthetic food and agriculture start-up, takes issue with Ginsberg's tree of life: "Arguably, I would say 'Synthetica' is not a new kingdom but a separate and distinct entity. There is no evolutionary link in between those life forms and the only shared thing is the fact that they were rationally engineered." That is, the living systems Ginsberg classes under Synthetica, on Delebecque's view, share no (genetic) common ancestor that can be vertically traced via evolutionary descent. Therefore, he suggests instead "a new dimension to the tree of life as Synthetica is not linked to the other kingdom through evolution but through a *rational choice*," where the "choice" is that of the synthetic biologist who wants, for example, "a yeast able to produce drug X"; "the pathway does not exist in nature but if I pick gene a from this archea and gene b from this bacteria it will work."[52] Indeed, after conversations with Drew Endy, Ginsberg revised her original diagram, offering instead a three-dimensional rendering of Woese's original genealogy (fig. 2.5). In it, Synthetica hovers above the other three domains, with lines tracing downward to connect, laterally, Synthetica to Bacteria, Archaea, and Eucarya. In this model, Synthetica is not an addition to the three domains found in nature; rather, it at once combines and is beyond them.

Biological trees of life, already modeled on family trees diagramming lines of inheritance and succession, are also theories of relatedness premised on biogenetic idioms. What Delebecque's critique of Ginsberg's

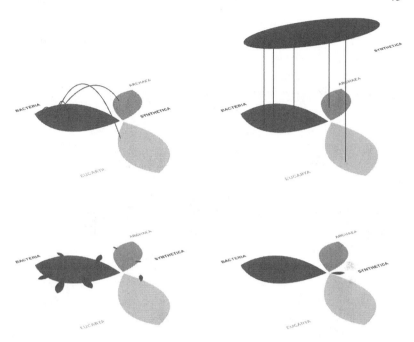

FIGURE 2.5. Alexandra Daisy Ginsberg, "The Synthetic Kingdom: Iterations," 2009. Courtesy of the artist.

"Synthetic Kingdom" (and her subsequent revision of it) points toward is that synthetic biologists view their work as undermining any neat bioge-netic lineage, whether evolutionary, cladistic, or otherwise. While Gins-berg's first tree tracks lineage understood genetically, Delebecque claims that "rational choice" is a better genealogical rationale than bearing ge-netic material in common. In tracing the relations of Synthetica to other kingdoms, he forwards "rational choice" as a justification for linking en-gineered living systems to their creators. Such a claim recalls arguments over how to define queer parentage, especially when parents may have neither legal nor biogenetic connections to their offspring.[53] In such cases, kinship is maintained and acknowledged by parents via an ongoing "vol-untary" choice to parent.

Anthropologists studying gay and lesbian kinship show how queer so-cialities undo naturalized versions of human relatedness by emphasizing sociality, exchange, support, and care rather than shared origins, geneal-ogy, or descent defined by blood. Judith Butler puts it like this: "the resig-nification of the family . . . is not vain or useless imitation, but the social and discursive building of community, a community that binds, cares, and teaches, that shelters and enables."[54] Kath Weston studied LGBT "chosen" families in the United States, demonstrating that they complicate traditional theories of kinship even as they are often modeled on affinal relations: "the category 'families we choose' incorporates the meaningful *difference* that is the product of choice and biology as two relationally defined terms."[55]

Voluntary kinship is not built on blood or marriage but rather is made up of choices to provide material and emotional support.[56] In such relationships, "biological ties are decentered and choice . . . becomes the defining feature of kin relationships."[57] Queer "families, in this sense, are not the neat nuclear units they were 'traditionally' believed to be . . . but instead groups of people who provide support and care for one another, and participate in the exchange of goods and services."[58] In so doing, queer kinship reveals the constructedness of all kinship systems, by "mak[ing] explicit the fact that there was always a choice as to whether or not biology is made the foundation of relationships."[59] By emphasizing "rational choice" over evolution or consanguinity, synthetic biological taxonomies are akin to what Weston terms "families we choose."[60]

The cells manufactured by synthetic biologists who build microbial pathways into transgenic yeast undermine and undo current models of taxonomy as they have been understood—that is, as joining together biological organisms related genetically by lineage or descent. Such cells couple diverse genetic parts, mixing together that which is seemingly unrelated and entangling genetic fragments, and in so doing, they persuade synthetic biologists that even putatively "natural" genealogies and taxonomies are similarly fashioned. Delebecque's claim that "rational choice" is a taxonomic rationale for synthetic biology echoes anthropologists' definitions of queer kinship based not on biogenetic connectedness but instead on ongoing "voluntary" design choices.

If kinship theory was first modeled on biology, as Schneider established, then queer or voluntary kinship demonstrated the constructedness of blood-based definitions of kinship, and finally, biotechnology, like queer kinship, problematized the "biological" as ever being natural. As Cori Hayden puts it, queer kinship reveals that "straight, blood-based kinship is itself a construction."[61] Now queer kinship theory may be applied to objects of synthetic biology—and perhaps all biotechnically made transgenic organisms—to make sense of how synthetic biologists arrange and define biological relatedness. Some social scientists have called for their colleagues to maneuver "athwart theory," "tacking back and forth between seeing theories as explanatory tools and taking them as phenomena to be examined."[62] Reading synthetic biologists' descriptions of relatedness alongside queer kinship literature reveals the extent to which emergent practices are complexly complicit with theory, as well as how the empirical and the analytic are mutually implicated.[63] This is not intended as a normative claim, and my stance is not celebratory. Rather than theorizing the empirical, I have here tracked ways in which synthetic biologists operate with and make sense of their own work using modes of reasoning I identify as already analytic.[64]

Conclusion: Queering Biocapital

Kinship is closely related to cultural understandings of property and how it is distributed—whether in the form of people as "goods" transferred via marriage or how property is passed across generations via gendered and blood-based definitions of lineal descent. Kinship theory recently has also made an important contribution to how scholars understand another newly vexing issue: that of biological property in biotechnology. In explicating her scholarly trajectory from analyzing lesbian kinship to bioprospecting, Cori Hayden makes this link explicit. She argues that "kinship and property have always been about the allocation of rights, responsibilities, inclusions, and exclusions"[65] and demonstrates historical linkages between kinship theory and intellectual property law: both have, she writes, "been based on the historically malleable but highly consequential line between Lockean nature, or raw material, and human enterprise, or culture. As is kinship, property rights are figured, in Locke's idiom, as a hybridized domain—the appropriation of raw nature and its mixture with human labor or artifice. But kinship and intellectual property are not just *like* each other. . . . Kinship and intellectual property have long been models *for each other*."[66] So too, Donna Haraway inaugurated kinship as a way to address "the question of taxonomy, category, and the natural status of artificial entities,"[67] arguing that in the products of twentieth-century biotechnology, "natural" sorts of kinship are rendered unnatural by being copyrighted, trademarked, patented, and branded. And yet, she finds, patented organisms are naturalized, as kin and kind become indistinguishable from brand.[68]

Microbes that manufacture antimalarials and diesel are multiply patented, and their genetic exchange, the "promiscuous" couplings of yeast,

bacteria, and wormwood, in the interest of solving the promiscuous merg-
ers of *Plasmodium*, *Anopheles*, and the global impoverished, is the source
of these microbes' proprietary value. Prior accounts of biotechnical property emphasize how species identity marks a biological object as having
market value: "biological kinds—taxonomic species—become a form of
biowealth, or brands."[69] In synthetic biology, however, it is instead the *loss*
of any meaningful and "pure" taxonomic identity that is the source of synthetic biowealth. Their very queerness, the transgressive entanglement of
diverse genetic material, is what makes synthetic microbes and transgenic
parts *worth* patenting or copyrighting in the first place. The next chapter,
with which this one is paired, extends this argument. It examines how dif-
fering proprietary regimes (such as patent and copyright) are built into
the objects of synthetic biology and the social, legal, and moral norms of
material exchange that synthetic biologists have attempted to install in
their field.

"To Make an Eye, a Hair, a Leaf"

syn·thet·ic (sĭn-thĕt'ĭk)
adj.
(In most senses opposed to GENUINE *adj.*)

3. Of a substance; an object of chemical invention or manufacture; *esp.* commercial products imitative of naturally-occurring materials (e.g., dyes); produced in a laboratory.

   1912 TH. CURTIUS *Berichte der Deutschen Chem. Gesellsch.* Synthetic the Coffee, Synthetic the Wine / Synthetic the Milk and the Butter Gloss Shine / On top of it all, even beer is not pure / Natural nutrition, you won't find to be sure. / Then do let the devil take it for free / That wretched synthetic-made Chemistry.

In the mid-eighteenth century, "synthetic" denoted ways of doing science, and chemistry in particular. The *Oxford English Dictionary* cites the first known use of "synthetic" to describe the synthesis "of organic compounds, produced by artificial synthesis" in 1753. By 1806 "synthetic" could be used (counterintuitively) to describe a mode of *inductive* experimental thinking, as, for example, in the use of the term "synthetic proof," which refers to scientific knowledge arrived at or proven empirically, by synthesizing new chemical compounds (*Oxford English Dictionary*). Here, scientific knowledge about the natural world is gleaned from making artificial things never before in existence. This sense distinguished synthetic chemistry from its earlier analytic (experimental) method.

   For synthetic chemistry, especially as it was promoted by Marcellin Berthelot in the 1860s and 1870s, "synthesis was a tool for knowing the world, a means of penetrating the secrets of the composition of natural substances."[1] Synthetic chemists of the time pursued "the double nature of synthesis as a practical as well as a theoretical activity," by which synthetic compounds would simultaneously illuminate chemists as to the properties of natural substances while also being useful as things in themselves.[2]

Berthelot claimed that chemical laws could be better understood if chemists composed increasingly complex organic compounds, proceeding from elements to small molecules to larger molecules. Still, the relation of the production of things to the production of theories is certainly not unidirectional. Novel theories may precipitate out of such synthetic methods, whether chemical or biological, but the things made also incorporate and make manifest long-held theories.

Nineteenth-century chemistry was "a tool for knowing the world," perhaps, but it was also a powerful instrument with which to intervene in it, and to do so in ways that were as fiscally as they were epistemically lucrative. The synthesis of aniline purple—otherwise known as mauve—by William Henry Perkins in 1856 ushered in the synthetic dye industry, in which chemists concocted riotous and vibrant colors entirely divorced from their natural sources. After mauve, fuchsia bloomed: in 1859 a Lyonnais chemist working for the silk-dyeing firm Renard Frères synthesized aniline red (or fuchsine), and the firm established the first patent on a synthetic dye soon thereafter. These first colors were soon complemented by Paris Violet, Bismarck Brown, and Congo Red.

The burgeoning synthetic dye industry, especially in Germany, shaped European patent law as competing chemical firms argued over what counted as invention and what was the proper object of patents: processes or products (especially when a single chemical process could produce hundreds of different colors). The German Patent Law of 1877 responded with "a relevant stipulation for chemical inventions," namely, "the fact that the law excluded the protection of chemical substances; a chemical invention would be patentable insofar as it concerned 'a particular process' . . . for the manufacture of such substances."[3] Synthetic chemists may have sought to better understand chemistry by synthesizing it, but some of the resulting compounds were so lucrative as to demand new legal structures to parse their ownership. Chemical processes, once named after their inventors, became the property of large-scale chemical firms. New ways of making new hues were now corporate property. Synthetic biologists, 150 years later, also aim to understand the living world by remaking it. They too worry over the legal status of the objects they manufacture. As they grapple with the "double nature of synthesis" as both practical and theoretical, the sorts of legal regimes into which synthetic biology might be slotted are rendered complex, debatable, and mutable.

Beyond such proprietary concerns, however, synthesis, in nineteenth-century chemistry as in contemporary biology, "raised the problem of boundaries. Did the chemist have the power to make life?"[4] The synthesis of urea by Friedrich Wöhler in 1828 is popularly assumed by many (even to this day) to have been the *experimentum crucis* by which vitalism was disproven by experimentally bridging the divide between organic and inorganic matter. In fact, it hardly made a dent in vitalist thinking. Allied chemists—Jöns Jacob Berzelius, Justus von Liebig—simultaneously attempted to synthesize organic compounds while admitting that such efforts would do little to dispel belief in a vital force. Liebig, a friend of Wöhler's, simultaneously predicted the imminent synthesis of organic chemicals like sugar and morphine while dismissing the possibility of making living things: "Never will chemistry be able to make an eye, a hair, a leaf."[5]

Perhaps not leaves themselves, but chemists now could make many of the chemicals once made only in leaves. Indigo was chemically synthesized in 1880, derived from o-nitrocinnamic acid rather than the flowering plants of genus *Indigofera* named for their color. First marketed in 1897, by 1913 most indigo was synthesized rather than natural, and the British indigo industry in India had collapsed.[6] Ignoring the receding horizon of synthetic chemistry's limits, all sorts of things that had previously been naturally sourced were, by the twentieth century, manufactured in a laboratory. "Synthetic chemistry provided many variants of dyestuffs and traditional remedies, from which effective compounds could be empirically selected. One can find similar shifts to 'construction' . . . in *avant-garde* art, with the move from analytical to synthetic Cubism, or in Kandinsky's abstractions, and then more obviously in Constructivism and Futurism. By recognizing the relative novelty of synthesis in this period, and the dialectic with new forms of analysis, we may open more general ways of understanding modernism."[7] Synthetic chemistry exemplified modernity's constructivist impulse, bleeding into fields from engineering to architecture to art.

Synthetic chemistry's aims and limits bear much in common with synthetic biology in that "synthesis"—the constructive or combinative assembly of new complexes from component parts—serves an inductive (or analytic) end that nonetheless saturates legal and proprietary problematics as well as philosophical concerns. By making new synthetic biological parts, fabricating functional genetic units, or engineering viable synthetic organisms in the laboratory, synthetic biologists, like synthetic chemists before them, seek to prove long-held theories gathered from discovery science. Some synthetic biologists even draw an explicit comparison between their work in the present day and mid- to late nineteenth-century synthetic chemistry. They posit that just as "fundamental theories of chemical structure developed concurrently with the explosion of synthesis" in synthetic chemistry, now "the convergence of analytical and synthetic approaches seems to be replaying itself in modern biology."[8] They further claim that material developments triggered theoretical breakthroughs in chemical knowledge. Such an analogy sheds light on how synthetic biologists themselves perceive their aims. The synthesis of urea, acetic acid, methanol, and ethanol, they write, impacted theories of atomic bonding and valences (as well as theories of life); the synthesis of benzene contributed to understandings of three-dimensional chemical structures. In her comparison of nineteenth-century synthetic chemistry to twenty-first-century synthetic biology, historian Bernadette Bensaude-Vincent asks whether "the adjective 'synthetic' has brought about the repetition of a past scenario. Is history repeating itself under the banner of synthesis?"[9]

CHAPTER THREE

The Rebirth of the Author: New Life
in Legal and Economic Circuits

Books are not absolutely dead things; they have a potency of life in them. . . . But then they are more than living; a good book is the precious life-blood of a master-spirit, embalmed and treasured up on purpose to a life beyond life. —John Milton, *Areopagitica*, xxxii

With each Property goes a life script. —William Burroughs, *Soft Machine*, 154

I'm writing this at sundown on September 13, 2013. Right now, my family is gathering in another city, sitting down to eat a ceremonial meal before attending synagogue for Kol Nidre, the religious service that inaugurates Yom Kippur, the Jewish Day of Atonement. As a child, I forced myself awake on this night because I had learned that while I slept, God would decide into which book I would be inscribed and sealed—the Book of Life or the Book of Death. If I stayed up, perhaps I could cheat the system. The Book of Life, I imagined, was a heavy volume over which a punishing God inclined his head, recording names late into the night as everyone around me slept. The next morning, bleary-eyed and thirsty, I would meditate on whether I was safe for another year, because I had either made the cut or gamed the rules by shaking off sleep. Having been reared on the books of Exodus and Isaiah, I never examined this seemingly ubiquitous association of books with life until I began talking to synthetic biologists.

That life is like a book or, more properly, a text is similarly unremarkable to most biologists working in the late twentieth century. The American life sciences have for sixty years marinated in a soup of linguistics, cryptanalysis, cybernetics, computational biology, and a heavy dose of Judeo-Christian thinking.[1] The metaphor now hangs threadbare, unexamined while omnipresent in American molecular biology, though in the mouths of biologists, the religious fervor and flavor of life-as-book does not always dissolve into secularism.

Molecular biologists in the 1950s and 1960s sought to "crack" or decrypt the "master program" of nucleotides that would "code" for amino acids. As historian of science Lily Kay reminds us, the description of genetic material as "code" preceded the association of "code" with computer binary or source code: it first loosely mapped out "a multiplicity of significations, definitional slippages, shifting meanings, and aporias" drawn by biologists in the 1950s from cryptanalysis and linguistics. Only later was the word inflected by usages in information theory and cybernetics.[2]

Later, as genomics ramped up in the 1980s and 1990s, the National Science Foundation, National Institutes of Health, and Department of Energy would fund the Human Genome Project, which sought to "read" the entirety of the human genome. In June 2000 newspaper headlines proclaimed that scientists had finished "reading the Book of Life: 'the most important, most wondrous map ever produced' ";[3] that "we are learning [the] language in which God created life";[4] that we had walked "on the path of biology's Holy Grail."[5] Politicians, scientists, and journalists likened the genome again and again to a mystical and numinously charged "Book of Life."

By the time synthetic biologists joined the bioengineering scene in the early 2000s, life was already resolutely bookish. In the early days of synthetic biology, Tom Knight put it succinctly: "The [genetic] code is 3.6 billion years old. It's time for a rewrite."[6] When testifying before the Presidential Commission for the Study of Bioethical Issues, Endy measured the speed of DNA synthesis in terms of works of literature: "20,000 characters" (where each character refers to both a letter of the alphabet and a nucleotide) is within the order of magnitude of Lincoln's Gettysburg Address (1,500 characters) or an average *New York Times* editorial, while "8 million characters" might encode the play "No Exit," he said, as well as Alice Walker's *The Color Purple* and Tolstoy's *War and Peace.*[7]

Though the metaphor by which synthetic biologists compare transgenic relatedness to queer kinship is implicit, they explicitly and regularly liken DNA to text. What often drops out of alignments of life with books, however, is that books are material things subject to legal restrictions on their uses. They are bought and sold, offered as gifts, lent to friends and colleagues. Readers are "free to carry [a book] around and to lend it to others. You are not free, however—beyond certain legal limits—to reproduce its contents in your own right for commercial gain."[8] Indeed, if we begin by taking biologists—synthetic biologists among them—at their

word that life is bookish, then such an analogy presupposes a constellation of legal and social rules particular to books, which in turn under-
writes how living things get commodified, used, transferred, copied, and reproduced within the American bioeconomy.

It is therefore no surprise that when they sought legal forms to regulate the growing economy of their manufactured genetic parts, the legal
arrangements they considered included not only patenting, which had
reigned in biotechnology since its inception in the mid-1970s, but a gamut
of intellectual property options—the public domain, a vague assemblage
of sui generis arrangements, and licensing (either by contract or by a
copyright-grounded assertion of rights).

Certainly, synthetic biologists are not the only ones entangled in these
questions—intellectual property is perennially interrogated by new technologies. Music distribution is one common example, but I am also aware,
as faculty at a university at the forefront of the massive online open courses
(MOOCs) trend, that such technological platforms also query assumptions about intellectual property. And I am not alone in my frequent and
mundane collisions with intellectual property law: my Shenzhen-made
iPhone clone is as small as a book of matches; I received a cease-and-
desist letter from HBO after downloading season 5 of *Game of Thrones*; a
departmental administrator asked last semester whether I was freely distributing to my students too much text by Jürgen Habermas. I didn't pay
for the music I'm listening to right now, and you probably didn't, either.
Synthetic biologists' worries over intellectual property are extensive with
and symptomatic of much larger cultural issues of the moment.

Taking a cultural-historical approach to the law that queries the status
of categories such as "property,"[9] this chapter traces how two conflicting
rationales played out in debates over intellectual property that were waged
in the first decade of synthetic biology. That "life is not what it used to be"
is as much about the sociolegal norms of biological practice as it is about
the epistemic content of life—indeed, principles of intellectual property
were often built into genetic substance. And the presence of both copyright and patent in synthetic biology is enabled by assumptions of what life
is like as well as how it *should be* remade, debugged, rewired, or rewritten.

A Lesson in Licensing Life

In 2008 I registered for Introduction to Synthetic Biology. Submitting
my course list to the MIT registrar, I was uneasy about diving headfirst

into graduate-level coursework in bioengineering. I hadn't taken a biology course since I was an undergraduate, and while I had spent two years working with synthetic biologists, I wasn't sure I would be able to keep up with the coursework. Still, I wanted to be more conversant in the principles of synthetic biology, experience how new students were inculcated into the field, and try my hand at some basic lab work.

A few weeks into the semester, I settled into my seat at the beginning of the three-hour class, turning on my laptop and taking out my notebook. Drew Endy wordlessly handed me a sheet of paper. I noticed him distributing similar sheets to several other students in the class as they also sat down. Across the top of the printed page, in block white bold print against a red background, was blazoned "PATENT." Beneath it, a pink rectangle labeled "PCHB" bore the following description: "The best gene encoding the super duper PCHB enzyme. Patent #1234567890." Along the bottom of the page were four strips of perforated paper, similar to the detachable contact information one might find on a flyer affixed to a community bulletin board. Each of the four strips listed the enzyme's name, as well as blank lines where I could fill out the following information: licensor, licensee, license fee (fig. 3.1).

After summarizing how US patent law applies to genetic material, Endy and Natalie Kuldell, the course's instructors, explained the goal of the day's in-class assignment: some students would act as inventors and patent holders (I was one of these, the inventor of a "PCHB Enzyme"), and others would behave like investors seeking to license the complete set of genetic parts necessary for encoding the *Eau d'E. coli* system (namely, six enzymes, plus an inverter and a terminator, one weak and one strong ribosome binding site, and a stationary phase PoPS source and a constitutive PoPS source).[10] All the investors received a sheet with a list of each of these parts, and they had to approach multiple inventors, negotiating with them for the use of their part.

The rules of the game were as follows: the first investor to license all necessary parts would be paid $100 cash by Endy. That investor would then need to repay any inventor who licensed her part at a fee. If the licensing fees exceeded $100, the student would have to make up the difference out of pocket, again in real money. In other words, if an investor did not secure a license to all necessary parts, he would walk away with nothing. If, however, he secured all the parts, he would earn $100, minus the sum paid out to each of the inventors with whom he had secured a license.

Before students scrambled to begin flustered licensing negotiations, Endy wished everyone "Good luck and great profits!" The spectacle of a

FIGURE 3.1. Author's copy of the handout for the patenting exercise performed in an MIT graduate bioengineering course.

room full of bioengineers-in-training haggling over strips of paper didn't fit my expectations of an MIT graduate course. Shouldn't we be pipetting something? How did this exercise teach students to become synthetic biologists? Had Mammon so infused academic circles that students were being taught to value money and legal licensing above the scientific skills they were expected to hone while at MIT?

I assumed that most inventors would want to license their products at a fee, but I was curious to see how the game would play out if I, along with a few other like-minded inventors, would freely license parts to investors. I started tearing off licenses for my enzyme and offering them to other students for free. Despite my efforts, by the exercise's end, not a single investor had been able to secure all the necessary licenses. Indeed, I noted that the very structure of the exercise pushed people to demand money rather than, for example, place biological parts in the public domain. With a self-satisfied grin, Endy ostentatiously snapped the $100 bill before returning it to his billfold and depositing it in his back pocket. For the following week, students were asked to answer three questions: "1. Was it easy or hard to license parts? 2. What determined the value of a part? Did inventors tend to overvalue parts? Did investors tend to undervalue parts? 3. The challenge system contained 12 parts. What would happen if you needed to assemble a system that contained 100 parts? Or, 1,000 parts?"

The lesson being imparted was as evident as the fact that Endy expected never to part with his cash. Some stakeholders claim that patenting will encourage investment in biotechnology and drive innovation by incentivizing bioengineers to manufacture lucrative genetic parts. Endy and Kuldell assumed that their students were pro-patent (if they'd reflected on their stance at all), and they wanted them to think differently about patenting. They believed that intellectual property law would slow the pace of synthetic biology by producing "patent thickets" that would dissuade researchers from sharing potentially useful genetic parts with one another. They wanted their students to think the same.

Lessons in synthetic biology were always simultaneously lessons in bio-engineering *and* the legal norms synthetic biologists hoped to convey to their students and colleagues—learning to be an MIT synthetic biologist, I realized, meant learning to be skeptical of, or even outright hostile to, US patent law and the commoditization of living things. It meant learning to affix social notions of benefit sharing and credit to engineered biological substance. Indeed, engineering genetic parts naturalizes commodification, just as commodification assumes discrete and alienable parts: "in synthetic biology our ideas about appropriation may (albeit subtly and gradually) shape our ideas about the nature of living things."[11]

The definition of copyright as a legal right that is extended to an author and the definition of patent as a privilege secured by someone who invents something new shifted significantly from the late fifteenth century to the Second Industrial Revolution at the end of the nineteenth century—and continue to do so to this day. Here is the essence of patent: the beneficiaries of patents are people who have made a new, nonobvious, and practical physical thing. Inventing something that can be patented, the logic goes, requires skill, if not ingenuity, and the ingenuity protected by a patent inheres in that object's materiality rather than in its style. Copyright springs from the legal status of the author, which "was formed in England in the course of the eighteenth century in part through the blending of a Lockean discourse of property with the eighteenth-century discourse of original genius."[12] Further, functional works are excluded from copyright according to US case law. Practically, securing copyright is about four orders of magnitude less expensive in the United States than applying for patents, and while patents expire after twenty years, copyright lasts up to 120 years.

Thomas Jefferson articulated US patent law in 1793, establishing criteria governing patentability that have remained relatively stable over two centuries: to be eligible for a patent, an object must be novel and useful.[13] In 1850 the US Supreme Court added a third criterion: nonobviousness.[14] In 1889 the US commissioner of patents rejected an application to patent a fiber identified in pine tree needles; this so-called "product-of-nature doctrine" has affected biotechnology by excluding found natural objects from the purview of patents.[15] The advent of industrial plant breeding and seed distribution in the United States in the 1930s placed economic pressure on the product-of-nature doctrine, and asexually reproducing plants were granted limited protection in 1930 (sexually reproducing plants received similar protection in 1970).[16] Companies like DuPont, which historically hewed closer to the chemical industry than to bioengineering, led the initiative to extend patent law in the United States to include engineered single-celled organisms (*E. coli*, yeast) as patentable "products of human ingenuity" and handiwork rather than unpatentable "products of nature."

As biologists began using recombinant DNA in the early- to mid-1970s to genetically modify biological materials, the product-of-nature doctrine was undermined: when Ananda Chakrabarty engineered a bacterium that metabolized oil slicks, did the resulting critter count as a product of nature, or was it a new and useful composition of matter? The Supreme Court ruled that it was "not nature's handiwork, but his own; accordingly, it is patentable subject matter."[17] This decision became the cornerstone of the newly commercialized American biotech sector, with over two hundred applications on engineered microorganisms flooding the US Patent and Trademark Office by December of the same year.[18] In short, living things now could also be, in the eyes of the law, "compositions of matter."

Yet patents on living things trouble the language in which US patent law is articulated. Is a genetic sequence or a bit of tissue extracted from a body somehow transmuted into something "novel" or improved by dint of its extraction?[19] How can the utility of a newly engineered variety of rose be evaluated? And how might one decide whether a triploid Pacific oyster is "obvious" or not?[20] Despite such questions, the application of patent law to engineered organisms did not problematize distinctions between nature and artifice but rather *foreclosed* them. Whether an object was biological was no longer a difference that made a difference to the US Patent and Trademark Office following *Diamond v. Chakrabarty*. Notably, this premise also happens to be synthetic biology's foundational rhetoric—to quote Endy, "What's the difference between building a bridge and designing a genome?"[21]

Developing in the wake of this court decision, biotechnology was conditioned by patent law and the concomitant figure of the scientist as not only an experimenter, discoverer, or theorist but also an inventor, someone equally comfortable in a faculty meeting and a boardroom.[22] Between 1980 and 2000 industry financing of academic research increased from 3.9 to 7.2 percent of universities' research and development budgets; in the same span of time, the number of patents awarded to universities rose from 390 to 3,088.[23]

Nonetheless, while the convergence of biology and commerce continues to drive biotechnology, the influence of mechanical and electrical engineering and computer science on synthetic biology ensures that the field does not fit neatly into the commercialized biotechnology sector. There is no consensus among the synthetic biology community writ large about anything, especially not intellectual property. Many synthetic biologists, especially those working in the private or industrial sector, embrace patenting. Craig Venter is the most notorious example of the latter—and I heard him frequently used as the butt of MIT students' jokes. They gleefully referred to the J. Craig Venter Institute (JCVI) as "Microbesoft." In addition to JCVI, the Synthetic Genomics Institute, Amyris Biotechnologies, and the labs of many synthetic biologists trained on the West Coast are more likely to think about synthetic biology on a continuum with metabolic engineering and hence seek patents for their products.[24] Between 1990 and 2010, 1,195 patent applications relating to synthetic biology were filed, 45 percent of which were filed in the United States. Most of these patents were filed by the corporate sector, however, and fewer than a third originated in a university.[25]

But the field is by no means like-minded in this respect: some synthetic biologists outright resist the patenting conventions that biotechnology received from chemical engineering and industrial agriculture. This ethos was most noticeable among Boston-area synthetic biologists—in particular, among MIT students, faculty, and graduates. John Wilbanks, the director of Science Commons, introduced legal scholar James Boyle to Endy. Boyle consulted with Endy on how best to encourage "openness" within MIT's Registry of Standard Biological Parts; Endy described the resulting guidelines as a "legal limbo" in which synthetic biologists would freely share parts without the use of patents or material transfer agreements. In an act of conscious resistance to US patent law, many MIT synthetic biologists ignored all possible legal restrictions, crossing their fingers that they wouldn't get slapped with an infringement suit. Their attitude is most evident when comparing publication volume to patents filed: while MIT synthetic biologists published more peer-reviewed articles than any other organization (academic or corporate) worldwide except the University of California, Berkeley, MIT didn't even rank among the top twenty synthetic biology patent applicants during the same period. Berkeley, however, has applied for the most patents globally.[26]

The sorts of disciplinary histories that MIT synthetic biologists told themselves about themselves affected the legal framework they attempted to construct for their newborn field. While engineers (and, increasingly, biologists) were familiar with patent law, computer scientists were reared in a field that applied both patents and copyright (as well as copyleft; see below) to software code. The application of copyright law to computer science was grounded in an earlier metaphor—that digital code was analogous to text, and that programming computers therefore constituted a form of authorship that should be protected. As legal scholars Arti Rai and James Boyle note, "software—a machine made of words, a set of algorithmic instructions devoted to a particular function—seemed to fit neither the copyright nor the patent box. It was too functional for copyright, too close to a collection of algorithms and ideas for patent."[27] As such, since the 1960s software has been subject to a confounding mix of both patent and copyright.

Synthetic biologists' debates about intellectual property did not, however, fall neatly along disciplinary lines—synthetic biologists with PhDs in electrical engineering, for example, did not necessarily embrace patent law. On the contrary, this mélange of disciplinary precursors catalyzed broad debates about the appropriate legal forms for the fruits of synthetic biology. Copyright and patenting are paired, potent, and coextensive social arrangements. In synthetic biology, they temporarily coexisted in a state of legal, economic, and moral pluralism, joined under the twinned signs of biology as both a manufactured artifact and a metaphorical text.

Patent or Perish?

Because synthetic biology was, in its early years, a changeable field on the cusp of experimentation and engineering, much time was devoted to airy debates over intellectual property and ethics. Yet of more pressing concern to young MIT synthetic biologists were the more pedestrian issues of credit and attribution. Synthetic biology's ambiguous status betwixt engineering and discovery science raised practical, even urgent, questions for graduate students, who were unsure what counted as credible scholarly output in the burgeoning field. Nowhere was this tension more readily apparent than in debates over how and where to publish.

In October 2005, just three months after I had joined the Synthetic Biology Working Group, seven graduate students from Endy's and Knight's labs met over lunch with three professors (Endy and Kuldell from MIT's Department of Biological Engineering and Pamela Silver of Harvard Medical School's Department of Systems Biology) to discuss what counted as publishable research. On the agenda were two questions drafted in advance by the grad students: "1. What sort of publishing/reward structure makes sense for Synthetic Biology? 2. What do we want to see out of a modern journal in general?" We met in the same room in which the Endy lab held its weekly meetings, facing one another across a square table. The conversation veered from a practical discussion of the journals to which students might submit their work to something akin to a group therapy session in which students aired their worries about constituting the first generation of researchers in an uncertain field. I sympathized, as I was nursing similar worries about what it meant to earn a doctoral degree in a precarious interdisciplinary field.

These graduate students were apprehensive about how they would professionalize and face the looming job market. Thomas, for example, was an older graduate student who had worked at Microsoft for seven years before joining Endy's lab.[28] At the time, his research project to minimize and modularize the yeast genome ("Yeast 2.0") was in danger of being scooped by a rival laboratory at a prestigious East Coast research university. Publishing first would secure his authoritative claim to this research, but he didn't know where such work *could* be published. Most of the papers the lab was then drafting, he realized, were not science papers; they were engineering papers.

Reshma Shetty, then a graduate student in Knight's lab, had recently returned from India, where she had been teaching synthetic biology. She was similarly panicked about how her work would be received and what would even count as something for which she could take credit. The stakes felt especially high to her, because she was engineering BioBrick parts, all of which she freely uploaded to the Registry of Standard Biological Parts website and banked in its MIT freezer. It was unclear to her how her efforts would yield dividends (either financial or social) once she graduated and entered the job market. Unsettled, she murmured that the "worst-case scenario [is that] none of us gets credit" and "the field flounders [because the] founders can't get resources to further the research." Silver paraphrased the problem she heard Thomas and Reshma Shetty voicing: "Where can you publish just ideas?"[29]

Endy admitted that in academic discovery science, it's "a race to publish first." But engineers are not held to similar expectations. Instead, Endy said, the "race is to define the problem and get resources to work. . . . If you're building something, you want it to be beautiful and functional." Or, put otherwise, "You don't need to build the first bridge, but the best bridge." Thomas leaned back in his chair to stare at the ceiling before sniping back morosely, "But we're publishing in journals that want only the *first* bridges."

A patent may be awarded to the first object of its kind, given that novelty, utility, and nonobviousness are the criteria that define US patent law. Yet such thinking is nonsensical when speaking of copyright law, where identity is bound to the form and style of a work of art. There is no first or subsequent *Gravity's Rainbow* or Rothko's *No. 61*; they are unique objects that can be copied or duplicated but retain stylistic uniqueness and authenticity.[30] To speak of synthetic biologists as pursuing not the first, but the best, as Endy did in attempting to placate his students, locates their work closer to creative or artistic production than engineering and closer to aesthetics than utility. Endy's distinction between discovery science and engineering hinged on the difference between function versus beauty, first versus best.

Experiencing a frisson over *Kitzmiller v. Dover*, which was being argued during these months, MIT synthetic biologists thought about their work as a way of *designing* life.[31] Design, for them, turned comprehensibility into an aesthetic: it is an act of human ingenuity and intelligence bent toward composing useful things. Indeed, design has, since modernist functionalism, broadly been defined as the marriage of creative expression to utility—"form follows function," goes the axiom, such that the final object reveals human discernment and art. In the work of designing life, the creative capacities awarded copyright are indistinguishable from, even impossible to perform without, the ingenuity typically rewarded by patent.

Exchange and credit in science have been the subjects of sustained sociological interest since the origins of science studies in the 1970s. Warren Hagstrom analyzed scientific work as a gift economy in which scientists are disciplined, such that lectures and talks are freely "given," with the expectation that the scientific community reciprocates by bestowing prestige on the scientist.[32] Latour and Woolgar revised Hagstrom's account by examining the cycles by which grant money produces data, which produce social status, which garners further grant money, transmuting cash into cultural capital back into cash and so on.[33] However, these portrayals are premised upon a distinction between facts and artifacts that is inapplicable to scientific fields in which *theoria* and *practica* are hand in glove. In appraising credit and attribution in synthetic biology, Emma Frow notes that because of the field's hybridity, "a tension [is] emerging between future commercial value and more traditional academic measures of reward and recognition."[34] Scientific prestige and authorship may be fueled by citationality and attribution, but prestige is inalienable. In contrast, both patenting and copyright rely on alienating the products of scientific work from their makers so that they may circulate freely in a technoscientific economy.[35]

MIT synthetic biologists may have been making living things in the service of understanding biology better, but the gap between discovery science and engineering, beyond the legal precepts being negotiated for ownership and sharing of biological materials, posed real problems for students like Thomas and Shetty and young faculty such as Endy as they embarked on academic careers in epistemically and legally ambiguous territory. One solution they considered was rejecting patents altogether in favor of copyright or, more precisely, "copyleft."

Share Your Parts!

Some MIT and Harvard graduate students and faculty in synthetic biology raised the possibility of committing to Free/Libre and Open Source Software (FLOSS) approaches to software distribution, in which rather than patenting, programmers copyright software and then embed "sharealike" licenses within the copyright to allow for the free distribution, modification, and sharing of the software and any subsequent versions of it.[36] Much of that legal stance was enabled by the extended metaphor by which life functions like computer software. For example, when I once asked Endy why he was so passionate about open-sourcing biology, he responded, "I don't want wheat fields in 2100 to operate like Windows 95."

These licenses are described as "copyleft" because they invert the function of copyright: rather than allowing the author to maintain rights and privileges over subsequent derived works, they allow people to modify, distribute, and copy the work while foreclosing the possibility that users can copyright derivatives of the original (a possibility when a work is released into the public domain). Because there is as yet no legal precedent allowing synthetic biologists to copyright genetic material, this notion remains legally untested, despite the fact that the FLOSS ethos permeated the culture of the MIT Synthetic Biology Working Group, from idiomatic expressions to sartorial choices.

The BioBricks Foundation is a nonprofit organization founded by the Synthetic Biology Working Group in 2006 to establish a synthetic biology commons and maintain the Registry of Standard Biological Parts, a centralized clearinghouse for sending and receiving standard genetic sequences among synthetic biologists. From Tom Knight's initial infusion of 12 parts in 2002, over the next decade the registry mushroomed to over 5,100 BioBrick parts, which are stored both digitally (as coding sequences) and materially (as plasmids suspended in the registry's freezers). Biological databases such as the registry are not merely "incidental, technical or prosaic adjuncts to work on the biological" but rather "are also recombinatory hotspots where cross-validating life form/form of life entanglements play out."[37] The MIT registry's existence and use engenders the faith that biology is "naturally" Open Source or (as one synthetic biologist told me in conversation) that it "wants to be free."

These parts were built by synthetic biologists and are almost exclusively used by them—71 academic laboratories and 128 undergraduate student International Genetically Engineered Machine teams, as of 2013, have received physical genetic material from the database.[38] To borrow a phrase from FLOSS activists, these parts are "free" as in both speech and beer.[39]

MIT synthetic biologists' hackerly commitment to FLOSS was underwritten by BioBricks' putative standardization and synthetic biologists' sympathy with earlier free software movements. They worked alongside such FLOSS stalwarts as Hal Abelson, Richard Stallman, and Gerald Sussman. Openness, they believe, allows for an iterative process that will engender "better" biology, as researchers work together to tweak, debug, and improve upon a system. By this logic, standardization breeds openness, openness furthers standardization, and the combination leads to "better" biology.

The BioBricks Foundation plastered slogans on T-shirts, bumper stickers, and the signature lines of synthetic biologists' e-mails, enjoining everyone to either "Share your science" or "Share your parts!" OpenWetWare, a website started by MIT grad students and managed by the BioBricks Foundation, made laboratory notes, protocols, and data openly readable and editable online. The ways in which social relationships and accountabilities are written into the stuff of genetic material were made visually explicit in a bumper sticker an OpenWetWare founder gave me: the "weak" hydrogen bonds joining purines to pyrimidines in nucleic acids are represented as a ladder of handshakes, a ritual and embodied agreement or gesture of goodwill between two parties (fig. 3.2). Following legal forms, handshakes, while a "weaker bond" than contractual agreements, are nonetheless often legally enforceable.

The BioBricks Foundation's operating logic was as follows: if living parts could be copyrighted, then, like free software adherents, synthetic biologists could copyright their products in order to then freely license those parts for open distribution among other researchers. Recipients could in turn modify and then exchange the resulting altered genetic material, again under a license. Ironically, such a system for undermining proprietary software depends on property rights in order to work—as one Harvard graduate student described it to me, the legal system is a "dual-use" technology that could be creatively modified to undermine itself.

The terms of such licenses vary depending on who has drafted them. The first and best-known copyleft license is the GNU General Public License (GPL). This system renders the sharing economy recursive: in order to modify and distribute a product, whether software or genetic material, one first has to agree to freely license all resulting material. Licenses generate ever more licenses, and software begets more software, as genetic material proliferates in the hands of synthetic biologists.

The model in place to enable the distribution of BioBricks is based on copyleft licensing with a few meaningful differences. The BioBrick™ Public Agreement (BPA), unlike the GPL and most other copyleft licenses, is not strictly "viral" because users of the license do not need to license or submit genetic components combined with BioBrick parts registered under a BPA agreement.[40] As the BioBricks Foundation explains on its website, "The contract is pretty simple. At its heart, one person (whom we call

FIGURE 3.2. A sticker distributed by openwetware.org affixed to the cover of the author's field notebook.

the 'Contributor') makes an irrevocable promise not to assert any existing or future intellectual property rights over something against the other party to the contract (the 'User')."[41] These are, following Cori Hayden's analysis of contracts in bioprospecting, "chains of entitlement and access" by which "a host of political liabilities and property claims, accountabilities and social relationships are being actively written *into* routine scientific practices, tools, and objects."[42]

While the hope of installing copyleft licensing or similar schemes in synthetic biology rests upon an analogy of genetic material as source code, the homology between copyleft as a legal form and the ontology of synthetic biology does not build solely on visions of software. It also assumes that biology is putatively "naturally" generative. Drafters and early users of copyleft licenses in synthetic biology imagined these legal contracts "spontaneously" or "virally" proliferating alongside living things that are modified and distributed with a difference, multiplying both legal forms and life-forms.

Such lively generativity is tied to the ways in which life can be capitalized upon. The patenting of living things generates a "double fetishism" in which cells' value is naturalized by their being "infused with vitality because of the erasure of the labour and regulation that allow them to appear 'in themselves' in such places as laboratories and simultaneously imbued with life because of their origin in living things."[43] However, something altogether different is at work when synthetic biologists propose using contracts or licenses to "free" biology from possessive ownership.

Namely, biological fetishism arises in part from obscuring the labor of biological production, which thereby amplifies a commodity's exchange value. In divorcing biological exchange from biological capital, copyleft licenses and contracts require not an obfuscation of biological labor but a careful accounting of all biological modifications performed and a concomitant regulation of all future labor to be performed on the thing being licensed. Rather than a double fetishism, this is an *inverted* fetishism. Biological generativity is assumed in the act of "copying," yet synthetic biologists' labor is explicitly inscribed in the exchange of living materials rather than swept beneath the Marxian rug.

Conclusion: Regulating Facts and Artifacts

In 2010 scientists at the JCVI inserted a synthetic *Mycoplasma mycoides* genome into a *Mycoplasma capricolum* cell, reporting that the resulting patented "synthetic" bacterium had successfully reproduced in culture. In addition to coding genetic material, Venter's researchers inserted four watermarks bearing noncoding nucleic acid scrawls. Using a cipher that encoded nucleotides into English, JCVI scientists assigned either a Latin letter or an ASCII symbol to unique codons, allowing them to "inscribe" not only letters but also spaces and punctuation into the sequence.[44] The fourth watermark bore, in addition to the names of the forty-six scientists who contributed to the project and a secret code that directs the codebreaker to a URL and e-mail account, quotations by J. Robert Oppenheimer,[45] Richard Feynman,[46] and James Joyce. The latter flourish was a phrase culled from *A Portrait of the Artist as a Young Man*: "To live, to err, to fall, to triumph, to recreate life out of life."[47]

Venter soon received a letter from the Joyce estate, complaining that he had breached copyright. Venter countered that he considered the quotation to be under "fair use." Science bloggers salivated over the legal implications of a copyright infringement case in which literature was inscribed in DNA: Carl Zimmer wondered in *Discover* magazine, "Would Venter have to pay for every time his microbes multiplied? Millions of little acts of copyright infringement?"[48] When the copyright to *A Portrait of the Artist as a Young Man* expired on New Year's Day 2012, the issue was rendered moot.[49]

One of the principles of copyright is that a product of human creativity can be copied in different media without losing its form—thus, perhaps microbes can transcribe Joyce's text, constituting "millions of little acts of copyright infringement." The very notion of "expression," in fact, "originated in the bid to find a distinction between book and machine. It was that element of a book that required initial authorship but could be copied without mind."[50] An expression can be copied identically, while copying a machine can only be achieved with a difference. The nexus of copyright and patent law in synthetic biology is technologically capacitated—because sequencing and synthesis allow synthetic biologists to shuttle between DNA as physical material and sequenced code and back again, divisions between information and substance demand answers as to how such a mutable entity might be legally managed.

Such issues, never neatly resolvable, reveal the extent to which both patent and copyright are historically and locally specific categories. Distinctions between patent and copyright have been installed in Euro-American legal thinking since the 1880s and are shored up by legal decisions that reinforce or amend these categories when a new technology calls them into question. (For example, is software a text, a machine, or a machine made of text? Once US courts extended copyright law to software, the answer became the first option rather than the second or third.)

If molecular biologists once read the "Book of Life" in order to understand biology, synthetic biologists now set their sights on writing that book in order to *make* new biological things. Making undergirds knowing. Such thinking has had profound consequences for how synthetic biologists understand the rights, privileges, and responsibilities that might accrue from making new living things.

While the analogy of DNA to code (in all its multivalency) has been amply studied by historians and rhetoricians of science,[51] synthetic biology demonstrates how such analogies come to matter not just as ways of thinking about biology but as ways of *doing* biology, including questions of intellectual property, publishing, credit, and attribution. Particular legal regimes emerge and get hashed out on the basis of such thinking. On the one hand, if DNA is like a book or a text or a code or music or some other performative act for that matter, perhaps it can be copyrighted. On the other hand, if it is an object of human manufacture, as the products of biotechnology have been defined by US patent law since 1980, perhaps synthesized DNA should instead be patented.

Debates over intellectual property waged by synthetic biologists reveal that the life sciences are today perhaps the clearest example of how copyright and patent law are definitionally unstable, continuously reworked to manage new things. Following Alain Pottage's thinking about bioprospecting, "It is not just that the entitlements inscribed in objects shift and multiply as they move across diffracted legal topologies, but also that the movement of these objects has the effect of sharpening and multiplying the diffractions of these legal regimes."[52] Law is not an inert tool that restricts or enables scientific practice. It is as mutable and elastic as technoscientific change, such that law and science (here, American intellectual property law and synthetic biology) are complex, evolving, and emergent symptoms of one another's respective ontologies.

Synthetic biologists have self-consciously inherited proprietary models from electrical engineering, mechanical engineering, biotechnology, and software programming. Synthetic biology disrupts and sometimes fractures divisions between copyright and patent law, problematizing what counts as words or things, creative expressions or pragmatic tools, and identity as inhering in materiality or form. As Adrian Johns suggests, while copyright and patenting have historically been carved up as a division between the literary and the mechanical, these divisions are by no means "natural kinds." Thus, "a reformation of creative rights, responsibilities, and privileges" "might adopt as axiomatic the distinction between digital and analogue, for example, for it is arguable that the act of copying is distinct in the two realms. Or it could embrace a more radical form of reticulation, recognizing multiple categories—genetic, digital, algorithmic, inscribed, and more—rather than a binomial pair."[53]

A more complex constellation of collations might be inaugurated by the biological, as "genetic, digital, algorithmic, inscribed"[54] become pivotal biological capacities. Copyright governs the *factual*, patenting the *artifactual*, and synthetic biologists seek to make biological artifacts engender biological facts.

Much More than Human

syn·thet·ic (sĭn-thĕt′ĭk)
adj.
(In most senses opposed to REAL *adj.*)

4. *Music.* Sounds that are composed electronically, by combining sine waves into harmonic tones (as opposed to sounds originating from voices or instruments)
    a. *Fig.* Pertaining to music that is inauthentic, unnatural: see SYNTHESIS n. 1.

        1932 ALDOUS HUXLEY *Brave New World* In the synthetic music machine the soundtrack roll began to unwind. It was a trio for hyper-violin, super-cello and oboe-surrogate that now filled the air with its agreeable languor. Thirty or forty bars—and then, against the instrumental background, a much more than human voice began to warble.

Twentieth-century synthetic music unspooled alongside synthetic art— cubism, Italian futurist theater, Soviet revolutionary choreography— echoing the dissonances between reality and representation, assemblage and sampling. In the 1930s "the idea of a *synthetic sound*, of a sonic event whose origin was no longer a sounding instrument or human voice, but a graphic trace, had been conclusively transformed from an elusive theoretical fantasy dating back at least as far as Wolfgang von Kempelen's *Sprachmaschine* of 1791, into what was now a technical reality."[1] By the early 1950s electronic music was promoted by cultural critics such as Theodor Adorno,[2] though criticized by composers who worried that synthetic sounds were a "dehumanization of music" and lacking in subjectivity, creativity, or human sentiment.[3] The introduction of digital synthesizers into music composition by inventors such as Robert Moog, who applied integrated-circuit technology to the synthesizers that would later bear his name, signaled the advent of electronic sounds in musical culture.[4] However, not everyone at the vanguard of electronic music embraced the synthetic: Don Buchla, who invented his own synthesizer independent of Moog, debuted his "Buchla Music Box" in 1965. He was nonetheless "reluctant to call his device a synthesizer. For him the word synthesizer had (and still has) connotations of imitation, as in the word 'synthetic,' meaning rayon and other man-made fibers. He did not regard his new instrument as a vehicle to imitate or emulate the sounds of other instruments."[5]

Such ambivalence toward the synthetic ran deep. Peter Zinovieff, another synthesizer inventor, integrated sampling technology into his machines because "real sounds have got so much complexity that they're better than synthetic sounds."[6] Synthesizers were named synthesizers though, *not* because they produced sounds that were unnatural, but because they performed a mathematical function: they allowed musicians to sample and assemble—to "additively synthesize"—sine waves into harmonic tones. Synthetic music (like synthetic cubism before it) is synthetic in the sense of being composed or assembled, *not* as in fake. Nonetheless, falsity, fakery, and artificiality adhered to the word so closely that early proponents of synthetic music were uncomfortable referring to their machines as "synthesizers." Yet other musicians and engineers would embrace the synthetic because of its intimation of the technically assembled, the not-real, and the inhuman.

Mother Mallard's Portable Masterpiece Co., formed in 1969, collaborated with Moog and was the first entirely synthetic music ensemble.[7] The group influenced better-known German "krautrock" bands of the next decade, among them Tangerine Dream and Kraftwerk.

The latter leaned into the machinic aesthetic, squeezing sound from the beeps and bloops of calculator buttons, outfitting themselves in lab coats, and building concept albums around the cybernetic interface of life and computing, as in 1978's *The Man-Machine*. Speaking in 1976 to a journalist about an anticipated joint project with David Bowie, a musician who reveled in a masterfully polished queer artifice, Kraftwerk member Ralf Hütter said of their collaboration: "I think it's the same kind of research for a synthetic man. We were fascinated by the creation of that concept and we tried to develop it by ourselves."[8] The synthetic bled out of Kraftwerk's soundscapes, sounding out a way of apprehending the relationship of humans to machines and the possibilities for creative manufacture that such interfaces enabled.

Synthesizers could, in theory, make *anyone* a composer, absent years of musical training or the skill, discipline, or ability to play an instrument. The introduction of synthesizers into music composition and performance raised the specter of automated artistry—the work of art in the age of

FIGURE INT-4.1. A photograph from the booklet accompanying Kraftwerk's *The Man-Machine* (1978) depicts the band's automaton-doppelgängers busily synthesizing krautrock.

mechanical audio reproduction. Electronic music does not constitute a unique object, but it banks on mass-produced copies of copies of copies of copies.

Like synthetic music, so too synthetic biology: both are enabled by the transformation of a living phenomenon—a voice, a molecule—into its inscription. The technology of sequencing and synthesizing genetic material allows translation, (mass) reproduction, manipulation, and analysis. And in both cases, the effect destabilizes the original referent, producing ontological uncertainty about "organic" sound or life: "a technological doubt has been introduced into the indexical readability of recorded performance. At any point what one is hearing might be the product of the synthetic."[9] This Turing-esque ontological uncertainty about the object produced—synthetic life, synthetic sound—is matched by worries about the mode of its production, which portends mass production and concomitant deskilling of scientific or artistic craftwork.

One promise of synthetic biology, an unrealized yet orienting trope among researchers, is that synthesizing genetic components and assembling them into viable synthetic organisms could be a computationally driven and automated process. Rather than building one-off biological systems, they propose pairing standardization with laboratory "assembly

lines" and production facilities some term "BioFabs" (short for fabrication centers), which would outsource the design of living systems and send
digital genetic sequences to commercial synthesis companies for assembly.
Such frameworks, synthetic biologists hope, would scale up engineering
living systems, automating the processes of both design and manufacture
and turning biology, as some researchers put it, from a "vocation" to an
"avocation." Some synthetic biologists aim to make standardized genetic
components so easy to put together that biological engineering would no
longer be a practice gained by years of doctoral and postdoctoral training.
Synthetic living systems, in effect, could be things anyone can compose.
Yet others express anxiety that "deskilling" biology could render PhDs
obsolete. Synthetic biologists promise (or threaten, depending on your
perspective) to automate bioengineering, and such mass production is already enabled by the widespread use of "assembly lines," high-throughput
DNA sequencers, and robotics operated by workers in synthetic biology
companies. Both electronic music and synthetic biology are premised on
democratizing, mechanizing, and automating production, whether of life
or of sound.

CHAPTER FOUR

Biotechnical Agnosticism: Fragmented Life and Labor among the Machines

No longer does the worker insert a modified natural thing [*Naturgegenstand*] as middle link between the object and himself; rather, he inserts the process of nature, transformed into an industrial process, as a means between himself and inorganic nature, mastering it.
—Karl Marx, *Grundrisse*, 705

M y windows were rolled up to protect me from the oppressive July weather as I emerged from Boston's Big Dig, that monument to inept urban planning. I maneuvered my car, with me acoustically and climatically insulated inside it, across congested traffic lanes toward the Seaport District. On my left, crowds of tourists thronged the sidewalks, gawking at the USS Constitution, better known as "Old Ironsides," an eighteenth-century wood-hulled frigate built in a Boston shipyard and launched in 1797. Over two hundred years later, the week leading up to American Independence Day in 2012 was something of a festival for the American transportation and military industry; Boston Navy Week and Boston Harborfest were simultaneously in full effect, with members of the US Marine Corps and Navy strolling along Seaport Boulevard. A strange bit of Cold War history was also being revived, as "Operation Sail," a parade of windships inaugurated in 1964, idled in the inner harbor.

The gridlock afforded me the time to think about the laboratory to which I was headed, whose founders seek to "automate" bioengineering. I recalled an MIT postdoc who had told me, "Genetic engineering has gone through its initial development phase, maybe it has gone through a Renaissance where highly skilled craftsmen are able to design really interesting and fancy and useful and sometimes beautiful things, but right now genetic engineering is trying to go through its Industrial Revolution." By invoking a "Bioindustrial Revolution," he meant transforming bioengineering from artisanship, in which one organism is engineered at a time, to microbial mass production. His comment reminded me that in the midst of the Chartist debates of the Second Industrial Revolution, Marx commented, "At the same pace that mankind masters nature . . . all our invention and progress seem to result in endowing material forces with intellectual life, and in stultifying life into a material force."[1] Could the same sentiment, I wondered, drive synthetic biologists' professed Bioindustrial Revolution?

I turned toward Drydock Avenue, where I was visiting Ginkgo Bioworks, a synthetic biology start-up company whose founders, Barry Canton, Reshma Shetty, Austin Che, Jason Kelly, and Tom Knight, I had first met when we were all at MIT. A reporter from *Science News* who had recently written about his visit to Ginkgo described it as follows: "engineering new forms of life starts with setting up a biological assembly line, the living equivalent of a transportation innovation. Synthetic biologists aim to reinvent biology in the same way Henry Ford revolutionized automobile manufacturing. Instead of installing standardized spark plugs or carburetors as a car moves down the line, the scientists tuck brand-new biological parts into the body of a bacterium."[2]

Ginkgo is on the top floor of a retrofitted industrial warehouse overlooking the pier at the far end of Boston's Seaport District. This sector of South Boston has long been a hub for manufacture. Initially built in the mid-nineteenth century for the nearby seaport and naval shipyard, the warehouses that line Drydock Avenue have for decades been sites of programmable machine tool automation developed at MIT[3] (they formerly housed the Boston Army Base and the South Boston Naval Annex). Today, these warehouses harbor biotech companies and seafood distributors.

My visit felt like a homecoming—I had known four of the company's founders since we were all graduate students at MIT, and I felt a vicarious thrill as I took the elevator up to the top floor of the warehouse, noting the other tech start-ups and biomedical companies on the floors that I passed on my way to the top. Ginkgo's entryway was spare and slickly designed, a bank of clocks ticking off times at various global locales.

Walking past the entryway, I arrived at a split-level workspace occupying the majority of the floor. Climbing a few stairs, I faced dozens of computers arrayed in rows running perpendicular to a bank of brightly illuminated windows overlooking the harbor. Here, I found Shetty, Kelly, and the other company founders typing away at their computers. Shetty greeted me warmly, and I soon learned why they were engrossed in their screens. Pulling up a chair alongside her desktop computer, Shetty showed me the proprietary software they had developed, which listed the various genetic parts and reagents stored in Ginkgo's living archive. Pointing at the interface, she showed me how this bespoke modeling software allows her to design new genetic systems and track their function.

In some ways, this software is a logical extension of an earlier impulse: these are the same synthetic biologists who, as graduate students, were devoted to compiling BioBricks into the Registry of Standard Biological Parts and following principles of "intelligent design"—streamlining viral genomes and rendering them "rational" and efficient. Now the same synthetic biologists had set their sights on streamlining, not just the stuff of life, but the ways in which they went about doing so.

Shetty and I next walked down a ramp that led to three other rooms that looked like standard wet labs. This was the domain not of company founders but of its employees, who followed the protocols that Shetty, Kelly, and Canton had ordered. Built into the company's physical architecture was a clear restructuring of biological making and knowing: design was separated, physically and practically, from the labor of putting genetic material together and inserting it into living cells. This trend was driven home to me when Kelly bluntly told me, "We want [biological] engineers at a computer and *away* from the [lab] bench."

The unambiguous way in which the lab's building plan architecturally reflected one of the three central tenets of MIT synthetic biology—the *decoupling* of design from manufacture—made me curious. Synthetic biologists working in industry seek to make both microbes and laboratories that operate like factories to churn out biofuels and other chemical commodities. What management theories, corporate cultures, and labor practices do they install in their factories?

In this chapter I tour, with a crowd of bioengineers-turned-managers, two synthetic biology companies: Ginkgo Bioworks in Boston, Massachusetts, and Amyris Biotechnologies in Emeryville, California. In Ginkgo, design dictates manufacture. In contrast, at Amyris design is rendered obsolete as *scale* supersedes *skill*. Both companies exemplify how synthetic biologists not only have embedded technical principles of manufacture (such as standardization, decoupling, and abstraction) into genetic material but also have mapped such dicta onto the labor relations that underwrite mass production in late capitalism.

A Biological Assembly Line

Ginkgo Bioworks is, as the *Science News* journalist put it, "a biological assembly line, the living equivalent of a transportation innovation." Its founders physically designed it with this history of US manufacturing in mind, in particular the assembly lines of Fordism and the management practices of Toyota. They compare both the cells they engineer and the laboratories in which they work to factories and assembly lines (and in particular those that manufacture automobiles). In so doing, these synthetic biologists position themselves as the inheritors of a long history of labor relations, mass production, and consumerism in the United States. They also draw on management theories—some from Japan—that these realms of American manufacturing have come to employ.

This cell-as-factory metaphor is certainly not new, as cells have been likened to factories at least since Claude Bernard described cells in 1878 as "like the factories or the industrial establishments in an advanced society which provide the various members of this society with the means of clothing, heating, feeding, and lighting."[4] For over a century, life scientists who want to mechanize, engineer, and operationalize living things have deployed this metaphor: "the way was paved for attempts to manipulate the efficiency of the cell's operations, leading ultimately to the literalization of the cell factory metaphor by modern biotechnology."[5]

Synthetic biologists may draw comparisons to assembly lines and manufacturing plants like those of Ford and Toyota, but the political economy in which they do so is radically different.[6] The political and economic circumstances in the summer of 2012 were nothing like those that buoyed the heydays of the auto industry—particularly those that gave the name to Fordism and the just-in-time supply chains of Toyota. Ford's Model Ts were not just at the forefront of American mass production in 1900; they drove American consumerism after the post–World War II boom allowed average citizens to buy luxury items—Packards, Edsels, and other automobiles— for the first time.[7] So too, Toyota's production process answered the needs of a rapidly expanding Japanese economy, first selling its products abroad, then filling the needs of Japanese consumers during the economic bubble of the 1980s.[8] Amid preparations for the 2012 Independence Day, however, the United States was not in nearly as rosy a position as Detroit in the 1910s or Japan in the 1980s. During the summer months, the country was haltingly dragging itself out of a Great Recession. That summer I read *New York Times* headlines offering such spooked proclamations as "Lost in Recession," "The Go-Nowhere Generation," "Is This Really the Worst Economic Recovery since the Depression?," "Still Crawling Out of a Very Deep Hole," "The Human Disaster of Unemployment," and "The Economy Downshifts."[9]

Yet despite the recession, industrial synthetic biology remained flush with cash. John Melo, the president and CEO of Amyris, was formerly a senior manager at British Petroleum; in a decade the new company he ran had accumulated approximately $309 million in funding from private equity firms. Synthetic Genomics Institute, which Craig Venter founded in 2005 in order, among other things, to engineer algal biofuels, received funding from British Petroleum and a $600 million investment from Exxon Mobil in July 2009. Venter's company has alternately been managed by veterans from General Electric and the US Department of Energy. Other biofuels companies are linked to Big Oil and pharmaceutical corporations such as Shell, Merck, Pfizer, Schering-Plough, Bristol-Myers Squibb, Chevron, Cargill, Dow, and DuPont.[10] Private funding is matched by substantial federal initiatives, such as the US Congress's 2007 National Renewable Fuel Standard program, which promises taxpayers 36 billion gallons of biofuels by 2022.[11] A former biofuels corporate employee, speaking under condition of anonymity, intimated to me that these joint ventures between synthetic biology and energy companies are examples of "greenwashing." Large corporations toss a (comparatively) small amount of their total earnings toward funding biotech research in order to promote themselves as dedicated to environmentally friendly energy sources while continuing their extraction and manufacture as usual.

In his "Fragment on Machines," Marx scrutinizes how the presence of machines in factories dissociates manual from cognitive labor, producing what he terms an effect of "general intellect" in which "workers themselves are cast merely as [the machine's] conscious linkages," who tend it and keep it running smoothly and ceaselessly. While Marx wrote the "Fragment" while surveying the failed proletarian movements of the mid-nineteenth century, contemporary scholars argue that it is prescient and "[r]ecognizable as a portrait of what is now commonly termed an 'information society' or 'knowledge economy,' in which the entire intellectual resources of society, from shop-floor production teams, to university-industrial partnerships, to the regional 'innovation milieux' of microelectronic and biotechnology companies, [are] mobilized to produce the technological wonders of robotic factories, gene splicing, and global computer networks."[12] Automation is not unique to synthetic biology; it is ubiquitous in biotechnology, pharmaceuticals, and many other technoscientific markets. So too, science studies scholars recognize that reading and experimenting with Marx can be remarkably illuminating when appraising biomedicine.[13] Industrial synthetic biology is crucial to understanding contemporary developments in the biological sciences precisely because it is diametrically opposed to the discipline's initial raison d'être of joining making (*practica*) and knowing (*theoria*).[14]

Applications of Marxist theory to the information economy in the last thirty years have focused precisely on such divisions of labor, such as the ways in which the information economy has generated a globalized, flexible economy in which "living labor" is gradually replaced by mechanized processes, often in the form of shop floor machine labor. We have, critical theorists tell us, moved from an age of *mass* labor to one of *socialized* labor. Maurizio Lazzarato, following Gabriel Tarde, describes the information economy as splitting labor into "automatism" and "invention." Automatic work is repetitive, imitative, and mindless, while inventive labor is differential—the work, Lazzarato writes, of "genius," the romantic legal fiction that buttresses copyright as creative expression to the neglect of labor.[15]

Like Antonio Negri and Michael Hardt, Lazzarato claims that the working class is now engaged in what theorists term "immaterial labor," which is the "labor that produces the informational and cultural content of the commodity."[16] As the only autonomist Marxist who has addressed questions of "life," Lazzarato is a key interlocutor when appraising how automated labor modifies the bioeconomy. Much of his thinking on this topic was triggered by French automobile workers' strikes at the Peugeot factory in 1989, a conflict he considered to be a response to disingenuous post-Fordist managerial promises of worker dignity and collaborative input. Negri, whose writing was similarly influenced by Italian workers' strikes at Fiat car factories in the 1970s, describes the dissection of intellectual from manual labor instead as post-Fordist "mass intellectuality" in which work is "shot through and constituted by the continuous interweaving of technoscientific activity and the hard work or production of commodities . . . , by the increasingly intimate combination of the recomposition of times of labour and of forms of life."[17] Indeed, this was apparent to me when I visited workers in a bioengineering factory modeled on a car factory.

The Organism Is the Product

Ginkgo's company motto is "The organism is the product," which means that Ginkgo works on contract for companies seeking synthetic versions of pharmaceutical, fuel, flavoring, cosmetic, and perfume compounds. A company contracting with Ginkgo describes the chemical it needs, and Ginkgo scientists figure out how to engineer a microbe to produce that chemical; they then sell the resulting organism to the contracting company. Much of the more than $6 million start-up grant Ginkgo received from ARPA-E (the energy research arm of the Defense Advanced Research Projects Agency) bought high-throughput machinery and robots that together compose what Ginkgo employees refer to as the company's "assembly line." As the Ginkgo founders describe their business model, "We have invested heavily in robotic automation and have seen a significant reduction in error rates as by-hand operations are replaced with robotic processes. Our automated platform increases the throughput and reduces the turn-around time for building and testing new organisms."[18] Having decoupled design from manufacture, Ginkgo scientists apply bioengineering principles to microbial design, but they automate building synthetic microbes.

Synthetic biologists have compared synthetic biology to Fordist assembly-line production and Frederick Winslow Taylor's principles of "scientific management" since the field's inception, and that analogy animates Ginkgo's work flow.[19] MIT synthetic biologists, inspired by Fordism, proposed designing standard cellular "chassis" that would function as stripped-down cells to which different genetic functions or features could be added, making explicit the link between cellular and automotive chassis.[20]

While I chatted with Ginkgo employees over lunch later that day, our conversation was halted by a stunning display, visible through the lunchroom's two walls of windows: the Navy's Blue Angels jet squadron swooped in a delta formation in rehearsal for its Fourth of July performance. The discussion that was interrupted when the Blue Angels swooped ostentatiously close to the floor-to-ceiling windows concerned what employees referred to as the "Ginkgo Canon." Canton describes the canon as "those writings that heavily informed or influenced you as an engineer, scientist, professional, or human in the recent past." Company founders, senior scientists, and undergraduate interns batted around their suggestions. In addition to biology texts, such as the *Enzymology Primer for Recombinant DNA Technology*, employees recommended works of science fiction that had shaped their thinking as synthetic biologists. They nominated an episode of *Dr. Who* as well as—an overwhelming favorite—Isaac Asimov's *Foundation* series.

They also listed Jeffrey Liker's *The Toyota Way*, which is required reading for all new Ginkgo employees. Notably, *The Toyota Way* is also assigned to all new Amyris employees (about whom more soon), who receive a free copy along with John Oakland's *Statistical Process Control.* While some synthetic biologists may make reference to Fordism and the automated mass production model, others adhere to the "lean manufacturing" ethos first installed in Toyota's automobile factories and popularized by management types as the "Toyota Way" or "Six Sigma" manufacturing. Amyris, like Ginkgo, embraces Six Sigma manufacturing principles, which audit manufacture and work flow in order to control product variation. A banner above the entranceway to Amyris's research space shows a three-part cycle, arrows connecting each word: "try, fail, learn." This motto resembles Ginkgo's methodology, in which "design, build, test" is a tripartite series of commands employees perform in an iterative loop.[21]

An alternative to Progressive Era scientific management, the Toyota Way borrowed Japanese ideas about aesthetic economy and work flow to promote "lean production" based on continuous improvement and process-oriented approaches to refining and maximizing manufacture. The Six Sigma program, forged by Motorola in 1985 and popularized by General Electric in the 1990s, was named with reference to statistical quality control, and the goal of this approach is to produce fewer than 3.4 defects per million (i.e., 99.99966 percent free of defects). Unlike Fordism, in which workers' tasks were distilled into streamlined, mechanized, and modular actions, the Toyota Way and Six Sigma encourage a democratization of labor in which (in theory) any employee could contribute to the improvement of the factory's functioning.[22] However, political theorists of late capitalism and pro-labor authors observing the actual effects of such management styles have reached bleaker conclusions as to their effects on labor practices and the political economy of mass manufacture.[23]

After lunch, a staff research scientist at Ginkgo shows me how the company's manufacturing process is physically divided so that personnel working in different rooms are devoted to either designing new microbes, building them, or testing them. The "design team," as I have already described, comprises PhDs and company founders, who design new genetic components in a large open office banked by computers. The "build team" occupies a wet lab downstairs. This group is populated by short-term contract workers, often undergraduates seeking summer internships or one-year positions (these workers are, in the company's argot, termed Ginkgo "padawans," after the adolescent Jedi apprentices in *Star Wars*).

Francis, who received an undergraduate degree in engineering and is now studying bioengineering before applying to grad school, is one such padawan. He shows me how the automated "assembly line" works. He explains, "The idea is to have a highly efficient and modular system of automation that keeps track of the fine details in the [DNA assembly] process, so that the designers can spend more time thinking about higher-level design." Such an enforced division between craft labor and managerial labor is not unique to synthetic biology—indeed, the factories of the Second Industrial Revolution were similarly partitioned. Historian of science Simon Schaffer notes that automated machinery troubled critics unsure of the "*site* of intelligence": "There was thus an unresolved contradiction between stress on the subordination and thus mechanization of worker's intelligence and on the coordination and thus cerebration of their labour."[24] This tension led to automated machines being fetishized as "thinking."

The assembly line Francis shows me is fully automated: all reagents and materials are bar-coded, and the proprietary software program checks to make sure that any two genes are compatible with one another before they are inserted into a recipient microbe. The protocol even directs build team members to a particular box in a specific freezer where the necessary genetic material or reagent is stored. Little biological knowledge is required to perform these pro forma recipes for synthetic microbes, as all that "thinking" is already programmed into the system. A padawan can join between six and ten genetic parts in a single ("one pot") assembly reaction and then transform the resulting genetic system into a microbe for testing (fig. 4.1).

Francis says that this system works on its own "without you having to think or make a mistake. It's quite powerful in that a user who has had minimal training in molecular biology can effectively do DNA assemblies that [otherwise] might take one or more trained PhD-level scientists to do." When biological design and biological manufacture are decoupled, the "site of intelligence" is reticulated among designers and automated robotic systems, while manufacture is limited to managing automated robotic assembly.

Imitating the lean-manufacturing ethos forged in Japanese factories, synthetic biology as it is practiced at Ginkgo follows an iterative loop similar to that used by hardware and software engineers: design, build, test, repeat. Yet the actual work of *thinking* about biology happens only at the design stage. As another employee told me, "The problem with biology is it's inherently so slow. . . . Our idea is if we can have a sort of automated pipeline where the operator has to intervene at some points but in general

FIGURE 4.1. An intern learns to program the Tecan Freedom Evo robot at Ginkgo Bioworks. Note the scattered pipette tips in the foreground. Photo by Aaron Heuckroth.

things are controlled by our informatics system and these platforms, then we can get accurately and well-built strains coming out of this [build] end and then we can test them."

While the notion of a biological "assembly line" trades on fantasies of biological mass production, it is also freighted with the labor relations that enabled Taylorism and the principles espoused by scientific management. Philip Mirowski suggests that Taylorism is the endgame of capitalizing science: "The final destination of market reform is to let commercial considerations modularize, standardize, and spin off almost every aspect of the process of scientific research, and consequently erase all boundaries between professional and wage labor."[25] Continuity, speed, repetition, and standardization are esteemed, but the icon of modernist capitalist productivity, made possible by uniform, interchangeable parts (such as the standard BioBrick genetic parts that had been foundational to MIT synthetic biologists), also entails the fragmentation of labor.

An earlier, humanist, Marx would have described this as the alienation of worker labor, but such a reading does not sufficiently encompass the chain of value creation in industrial synthetic biology. Once biology itself is fractionated into so many standardized parts, it follows that the work of assembling new biological systems out of these parts is similarly fragmented, standardized, and repetitive.[26] This work also embraces aesthetics of speed and economies of scale, even to the point of overthrowing biological design in favor of an accumulative biological agnosticism.

Heather versus the Robot

The trend toward robotization, acceleration, and automation of biological work is not unique to synthetic biology, as historians of the life sciences have demonstrated for allied fields such as bioinformatics.[27] However, it is exacerbated by synthetic biology's project, which is *making* rather than, say, annotating, sequencing, or data-mining extant biological material. One would be hard-pressed to enter any well-equipped and appointed biomedical, biotech, or pharmaceutical laboratory these days without finding within its walls a single robot, if not a whole clutch of them.[28]

The seeds of such approaches to standardizing, automating, and accelerating laboratory work were already germinating in synthetic biology's early days roughly from 2001 to 2006. Biology, MIT synthetic biologists argue, is no longer a discovery science, nor is bioengineering "artisanal." Synthetic biologists have sought to achieve this goal by redesigning living things to be more "rational": they fractionate living things into component parts, they standardize those parts, and they work to abstract or insulate different levels of biological complexity from one another. "Decoupling" biological design from biological manufacture, a goal written into MIT synthetic biology's charter, means that a biologist can design a new genetic system and outsource its synthesis to a company. Synthetic biologists have tried, in short, to make cells operate like the microscopic factories they already believe them to be, and they similarly want industrial synthetic biology laboratories to have the same design features that they install in their synthetic cells. The automation and acceleration of biological labor entail and are premised upon the fractionation and standardization of biological commodities forged from living matter.

In 2006 Endy asked graduate students in the Synthetic Biology Working Group to prepare a video tour of the lab for the newly established Synthetic Biology Engineering Research Center (a DARPA-funded research initiative). In the video, directed by (and starring) the same students who would later found Ginkgo, lab work is choreographed in rhythm with Britney Spears's electropop 2003 single "Toxic." The video offers a snapshot of what daily work looked like at the Endy and Knight labs during that period. As in countless other molecular biology labs, students spin petri dishes, using a bent glass rod to evenly suspend bacterial cultures on the plates. Canton and Kelly sit facing away from one another at lab benches, repetitively pipetting samples; another grad student loads plastic tubes into a microcentrifuge.

The video's climax is a staged joco-serioso contest between woman and machine: Endy's research technician Heather Keller pipettes samples into a gel in preparation for electrophoresis. Meagan, a research technician working in Knight's lab, is also tasked with pipetting samples, but an automated liquid-handling robot aids her. She kicks back in a chair, her feet propped on the edge of the robot's platform. Stretching and yawning, she opens a book as the robot sets to work. The video jump-cuts back to the Endy lab, where Keller pipettes with feverish intensity; the film speeds up and her movements grow jittery and repetitive in time-lapse. She is no match for the robot, who tirelessly outpaces her. In the meantime, Meagan has put her head down on her desk to nap while the robot persists in its labors. Seven years later, the same students trained at MIT to have good "lab hands"[29] have now delegated those embodied tasks to the part-time workers and robots populating the "build team," while they themselves are seated at computer terminals surveying the wet lab beneath them.

Such workaday research anxieties about pipetting quickly enough are, in the industrial context, bound up with overwhelming worries about the speed, acceleration, intensity, and expense of multimillion-dollar research agendas. Industrial synthetic biologists declare—often proudly—that speed and scale have recently replaced comprehension. In a *Nature Biotechnology* review article, the vice president of research and development at LS9, a Bay Area biofuels company, put it this way: "When I think of synthetic biology, I think of all the things I used to do as a biologist, slowly, methodically, and not always with a lot of success. . . . Today, most of these methods can be automated such that robots can carry out thousands of experiments effectively, efficiently and with excellent success."[30] This rhetoric of speed is echoed by a venture capitalist who cofounded Flagship Ventures, which funds LS9: "The way I like to describe it is recapitulating about 30 years of *E. coli* engineering in about eighteen months."[31]

Others worry, however, that even with the automation of synthetic biology, biofuels research still cannot be fast enough. An article in *Meta-*

bolic Engineering bemoans the "extremely laborious, costly, and difficult" work synthetic biology demands, even as it seeks "to accelerate the
design-build-test loops" that emulate Toyota's lean manufacturing. The
authors report that "economical production of artemisinin in yeast [by
Amyris] has already accounted for over 150 person-years worth of work
and counting. . . . It took DuPont and Genencor approximately 15 years
and 575 person-years to develop and produce 1,3-propanediol and Amyris Biotechnologies about 4 years and between 130 and 575 person-years
to make farnesene."[32]

Surveying the Human Genome Project (HGP), science studies scholar
Michael Fortun comments that "speed becomes a useful trope for mapping the social, institutional, and conceptual reconfigurations being created in and through human genetics today."[33] Caught up in this rhetoric
of speed, industrial synthetic biologists must continuously scale up, in-
tensify, and accelerate their manufacture just to keep pace with their own
rhetoric.

"The Stupid Way Is the Best Way"

A month after my visit to Ginkgo, I fly to San Francisco to learn more
about West Coast synthetic biology. Upon entering Amyris's front lobby, I
am confronted by a massive minimalist wall sculpture of the Buddha—not
the portly grinning Chinese Budai but the serenely dour Southeast Asian
deity. I meditate on this effigy while I wait for a security guard to process
my entrance clearance, and the smooth contours of its countenance seem
to enjoin me to be moved by some no-mind Buddhist-lite revelation: perhaps that Amyris, best known for synthesizing a potent antimalarial drug,
seeks to repair ecological catastrophe by engineering interconnectedness
and compassion. Instead, I grumpily reflect on the Silicon Valley–chic vibe
the company's campus conspicuously advertises.

In many respects, Amyris looks radically different from Ginkgo. Whereas
Ginkgo is a small start-up company, Amyris employs hundreds of researchers and workers on a few floors of a modernist suite of offices in industrial
Emeryville, California. Ginkgo's ethos builds on electrical engineering approaches to standardizing and streamlining biological components, which
Endy and Knight inculcated into the company's founders while they were
students at MIT. Amyris is instead the legacy of large chemical engineering
companies like DuPont, which coax microbes to produce useful chemical

products, such as drugs and fuels. Founded by Jay Keasling in 2003, Amyris cut its teeth and earned its name engineering the aforementioned antimalarial compound artemisinin.[34] Because artemisinin is molecularly similar to several fuel precursors, the company has since turned its attention to getting yeast to manufacture biofuels in sufficient quantities to compete with fossil fuels.

Before my visit to company headquarters, I was already familiar with Amyris's high-throughput factory approach to synthetic biology, in part from hearing its staff scientists speak at synthetic biology conferences over the years but also because of the reputation Amyris has earned among synthetic biologists. A University of California, Berkeley, graduate student once warned me that he had heard Jack Newman, the chief scientific officer of Amyris, say, as he paraphrased, "instead of thinking about things, [Amyris] just makes all sorts of different variations of one object, screens them, and picks the best." The grad student who shared this with me while we talked in his basement laboratory was deeply unsettled by this approach to doing synthetic biology. He opined, "as a scientist, that freaks me out because the objective is to get the field to a point where you don't need to *understand* any of the biology. Instead you can just cram through an empirical problem you're trying to solve." When I pushed him to recall how Newman had formulated this notion, he paraphrased Amyris's business strategy as "the stupid way is the best way." This modus operandi is of a piece with the epistemology of "Big Data" sciences, in which economies of scale trump theorizing: knowledge is supposed to emerge spontaneously from the database, as long as it is big enough. However, it also distinguishes the deskilling under industrialization from that under bioindustrialization, in that not only manufacture but also design lose their "site of intelligence."

Jeff Ubersax, Amyris's associate director of high-throughput sequencing, is boyish and ebullient. I meet him in the lunchroom, which is designed to resemble a midcentury diner, outfitted with synthetic materials of the time: Formica countertops and red Naugahyde booths. Rich Hansen, the senior director of research operations, and Zach Serber, the director of biology, join us; all three men are young and dressed casually in jeans and hoodies. The company cultivates a Googlesque persona, offering employees free healthy lunches, massage breaks, on-site seminars, and yoga retreats. On sunny Friday afternoons, I found staff members barbecuing and drinking beer on the manicured lawns that the expansive lab windows overlook.

Before the three biologists-turned-managers give me a tour of the lab, we talk about Amyris's business strategy in one of the company's many corporate conference rooms. There, they explain to me that "putting DNA together [is] no longer time-consuming and it doesn't require expert practitioners." Serber begins by telling me that one of the fundamental differences between academic and industrial synthetic biology is that while the former standardizes *parts*, the latter standardizes *practices*. Rather than following Endy and Knight's vision of designing rationally engineered genomes to meet human-defined ends, Amyris scientists instead "found they're better off making tens of thousands of genotypes and looking among them for the one closest to the one that has the phenotype you want." Accordingly, Amyris makes lots of new organisms. Lots and lots of them. In fact, one of the company's many managerial slogans is that "progress is equal to the product of the number of things you try and the quality of the things you try."

Amyris's corporate strategy, I learn from them, is based on a proprietary synthesis technology called RYSE, an acronym for rapid yeast strain engineering. This technology puts modular ends on pieces of standardized DNA, assembling a frozen archive of thousands of parts called RABits (RYSE-associated bits). Ubersax compares RABits to MIT's BioBricks, explaining that BioBricks were too expensive to implement because each standardized part was flanked by a genetic prefix and suffix (which would have to be synthesized at a cost). RABits, on the other hand, are modular but can be assembled without restriction enzymes. Automated software then designs a protocol by which to stitch together a series of up to nine genes in a single one-pot reaction, forwarding that protocol directly to robots that then perform the genetic recombination. Amyris's library of parts, which is named the RABit Hutch, is massive: as of 2012 it exceeded 60 million base pairs, over 90 percent of which are exogenous genes from yeast genomes. The RABit Hutch was launched in January 2010 and, my three tour guides proudly tell me, on that date the number of strains made per unit time "shot up" precipitously. An automated computer system (named THUMPR) assembles genetic parts into combinations of around nine genes. By 2012 over five hundred genomic assemblies were performed at Amyris each week. None are made by human hands.

After our initial conversation, the four of us walk from the lab's entranceway into an interior hallway lined by laboratories. Ubersax says that historically, putting DNA together has been an "artisanal" process "with a high likelihood of failure," a condition Amyris has sought to eliminate.

Amyris scientists initially developed RYSE technology so that synthetic biology would be easy enough for people with undergraduate degrees to do. But, he tells me matter-of-factly, "Now we use robots instead." None of the design work associated with RYSE is done by employees working in wet labs—workers care for the robots that assemble, grow, and culture the associated yeast strains and maintain the stocks of fermenting synthetic yeast.

When I ask them whether they can design new yeast strains without setting foot in the laboratory, Serber, the director of biology, laughs—"I do it from airplanes," he assured me. Jack Newman, Amyris's chief scientific officer, told a reporter from the *Washington Post* a similar tale of post-Fordist telecommuting: "For scientist Jack Newman, creating a new life-form has become as simple as this: He types out a DNA sequence on his laptop, and clicks 'send.' And nearby in the laboratory, robotic arms start to mix together some compounds to produce the desired cells."[35]

In the next room over, in the DNA assembly suite, several dozen robots are picking cultures, moving them from plates to wells, and pipetting solutions to make new yeast strains. These robots are named after fictional pop-culture robots—arranged in one row I see Bender, Wall-E, and R2D2. The only thing absent from this massively high-throughput facility, I notice, are the scientists—the only biologists in the room are the three people giving me the tour.

I ask who is monitoring all these robots. Ubersax and his two colleagues hedge their answers, responding vaguely, "Amyris employees." After the tour we adjourn again to the conference room, where Serber tells me that my question is a "delicate one." I ask whether I had inadvertently stumbled onto a sensitive subject, and he says that a few years ago, the company had to "repurpose" its employees, a task that fell to managers such as him. The problem, as he explains it, was that the scientists managing the robots, who at the time were postdoctoral fellows or scientists with PhDs, were stigmatized by their colleagues. Bench researchers demeaned their work as being merely "technical" rather than scientific. Even professional biologists were referred to disparagingly as "technicians" because their work was widely considered by some to be "repetitive and mindless."

As a result, and perhaps unsurprisingly, the demographics of staff scientists at Amyris have rapidly changed over the last decade. At its inception in 2003, scientists, usually starting at the postdoctoral level, did most of the work. The manufacturing team was composed of scientists who had earned PhDs from nearby universities and were keen to get in on the ground level of a hyped new biotech firm. Soon, however, Amyris began hosting competitions to see who was better at yeast strain engineering— humans or machines. The humans, it turned out, made more mistakes and worked more slowly. The robots performed in the competition eighty to a hundred times more efficiently than their carbon-based counterparts. These scientists were, Ubersax says, "repurpos[ed] . . . toward new activities." In effect, synthetic manufacture of microbes was deskilled—the company transitioned from hiring postdocs to graduate students, then to people holding bachelor's degrees.

The increasing penetration of machines into the workforce divides employees into two roles with two radically different sets of expertise. The fragmentation and hierarchization of labor are exacerbated by the contemporary technoscientific, free-trade, and information economies, as Donna Haraway has stressed with regard to the labor politics of the semiconductor industries of Silicon Valley: "Deskilling is an old strategy newly applicable to formerly privileged workers." Such labor is feminized, whether performed by women or not, "able to be disassembled, reassembled, exploited as a reserve labour force; seen less as workers than as servers."[36] At Amyris, managerial and manual labor is similarly divided along lines of gender, class, and nationality: "While workers in the 'high' sector"—people like Ubersax, Serber, and Hansen—"may be technologically skilled and relatively secure, and perhaps may even identify with their work as part of 'the brains of the operation,' the 'low-end' service-sector worker is poorly paid, insecure, untrained, deskilled."[37]

Biology without People

A room full of gas chromatographs, the next stop on our tour, was again devoid of people (fig. 4.2). My tour guides tell me that just four years ago, this room had only three gas chromatographs, all operated by trained biologists. Serber now surveys the room in apparent satisfaction. Beaming at the gas chromatographs, he assures me that the current model is more efficient and yields more accurate data. Similarly, a room dedicated to colony-picking robots is run entirely by a single staff researcher, a petite middle-aged woman who hurries from one robot to the next, loading samples.

Colony picking is typically a time-intensive and repetitive laboratory practice in which a researcher uses a pipette to transfer single microbial

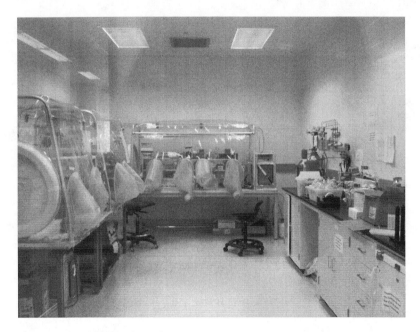

FIGURE 4.2. An empty lab bench at Amyris, August 2012. Synthetic microbes are growing in vats in the middle ground. Photo by author.

colonies from one plate to another, either for testing or for further analysis. The throbbing drone of the machines is thunderous, and my tour guides shout to make themselves heard above the noise. While Amyris once employed fleets of workers who picked microbial colonies, handpicking can cause repetitive-stress injuries. Robots now move strains into ninety-six-well plates, which are then transferred to another room, where automated "shakers," machines that agitate the cultures as they grow, ensure an equal distribution of oxygen and media to all developing microbes. In the room I visit, there are twenty-four shakers, named for comic book superheroes, that never idle. They tirelessly agitate synthetic yeast that grows and generates a hydrocarbon that can be used as a precursor for fuels and lubricants.

In developing and streamlining its research unit, Amyris explicitly modeled itself on the Joint Genome Institute in Walnut Creek, California, one of the central laboratories of the HGP. In the final months of the HGP, my guides tell me, many employees at the facility had suffered repetitive-strain injuries, the result of methodically peeling back and unsealing agar plates. At one point, the injuries were so bad that the laboratory shut down for a month while its employees physically recovered.

Eager to avoid a similar work stoppage, Amyris dismissed its colony pickers and bought more robots.

That Amyris models its work flow on a laboratory of the HGP is not coincidental. In the 1990s the HGP was responsible for importing the principles of Big Science—namely, large-scale, multilaboratory collaboration, masses of data, and expensive equipment—from physics into the life sciences.[38] Throughout, biologists critiqued the HGP for being ascientific, routinized, and dominated by data yet lacking in theory. Postdocs regularly quit, complaining that the work was boring and repetitive. In 2000 Craig Venter, then head of Celera Genomics, the for-profit company racing the National Institutes of Health–funded Human Genome Initiative,[39] offered a *New Yorker* reporter a tour of his laboratory space. The resulting description closely aligns with what I observed at Amyris (and also recalls the earlier journalistic description of Ginkgo). "We stopped and looked over a sea of machines. 'You're seeing Henry Ford's first assembly plant,' [Venter] said. 'What don't you see? People, right? There are three people working in this room. A year ago, this work would have taken one thousand to two thousand scientists. With this technology, we are literally coming out of the dark ages of biology.'"[40] This is one manifestation of the twenty-first-century bioeconomy: when biological manufacture is scaled up and routinized, biology no longer requires biologists.

The biologists who remained at Amyris decamped from the lab bench to middle management, where they are trained in a suite of skills for which they were not prepared in graduate school. At Amyris, biologists work with executive coaches and peer coaches to become middle managers, learning how, as Serber told me, "to move teams in a new direction with the minimum of angst and resistance." When laborers are gradually eliminated from post-Fordist production, such cutbacks have "paradoxically been accompanied with increasing managerial concern about the quality of the remaining workers."[41] If Taylorism and post-Fordism were examples of how management came to be scientized, the entrance of synthetic biology into market economies is, instead, an example of how science newly has come under the sway of management theory, turning biologists into group managers, collaborative problem solvers, and team players.

Conclusion

By the end of 2013, six months after my visit, Amyris reported building over 1,500 "new organisms daily." It has accumulated more than three million synthetic life-forms in the last decade, almost all of which are archived in the company's −80° Celsius freezers. It is, Amyris PR attests, the world's largest frozen zoo of synthetic organisms.[42] If none of these organisms were designed by synthetic biologists, what is a synthetic biologist becoming?

I received one answer from a high-ranking scientist at a large synthetic biology company. He told me that his company's business model was "high throughput"—the company would "synthesize lots of yeast, run lots of different variations in order to hit on something."[43] In terms of workforce, his company "wants biologists only doing design work." When I asked whether any biologists still worked in his wet lab, he unequivocally answered no. Biologists at his company should only be "designing things on computers and going to lots of meetings." He mused that within the decade, "no biologists will be needed for wet lab work," and they could "save money by hiring fewer people." In that event, they wouldn't "need to pay for health insurance and they [the robots] work 24/7."

Anthropologists and historians of science have recently adapted Marx to the life sciences, asking what theoretical hay might be made by appending the prefix bio- to *Capital*.[44] How, they ask, do biotechnology and bioengineering speculatively generate surplus and profit? The value of biotechnical products, as they ramify in globalized economies, is rhetorically and materially fastened to the imagined potency of biological substance. If living things are capitalized upon for profit, what is the work by which such things acquire value and become commodities? What are the means of production, the labor relations, and the technological infrastructures that undergird such profit-seeking enterprises?[45]

Understanding living things as raw material that can, through the application of labor, be transformed into a value-added commodity that circulates in capitalist markets, scholars have excavated the Marxian underpinnings of biological markets for stem cells, kidneys, and pharmaceuticals. I take synthetic biologists' comparisons of microbes to factories as an extension of an unstated comparison of organisms to laborers.[46] Describing the biological generation of value as something "innate" to living things, which "naturally" act like microscopic laborers or factories, Stefan Helmreich argues, obscures the work required to manufacture and care for synthetic life-forms.[47]

Fantasies of deskilled, scaled-up, and automated synthetic biological factories in which manual and intellectual labor are dissociated trade on concomitant imaginations of standardized and fractionated cellular factories in which biological information (sequences) can be divorced from biological substance (molecules). Theories of biocapital have, to date, relied primarily on the humanist Marx of *Capital*. However, reading contemporary bioeconomies through the lens of affective and immaterial labor affords a different view of how biological value is lately generated.

Immaterial labor extends beyond the factory, reticulating into every aspect of our daily experiences and lives. Yet as industrial synthetic biology makes apparent, the decoupling of design from manufacture generates a further rupture that dissociates making from knowing. This is a biocapitalism forged in economies of speed, scale, and theory's opposite—an agnostic materialism. Robert Proctor and Londa Schiebinger have identified "agnotology" as an epistemic effect,[48] yet such epistemologies are also installed into technoscientific products. Agnosticism is built into living stuff, both in synthetic biology and in other bioeconomies, and it has a market value. Under these conditions, conception and execution become separate activities: design and manufacture are decoupled at Ginkgo and Amyris, done by different people, in separate places. At the level of biological design, scale replaces skill.

In decoupling design from manufacture, industrial synthetic biology has *also* decoupled its central precept—that making furthers knowing. While synthetic biology once sought to make new life-forms in the service of understanding life better, biological making and biological knowledge, *practica* and *theoria*, are here separate domains. In an ironic turn, making living things, in such companies, no longer furthers biological knowledge. Instead, it is antithetical to it. Biomanufacture, at delirious speeds and mushrooming scales, produces millions of new life-forms, each of them the product of technoscientific agnosticism.

What Comes Before

syn·thet·ic (sĭn-thĕt'ĭk)
adj.
(In most senses opposed to ANALYTIC *adj.*)

5. *Sci., Philos.* Of or pertaining to various, mutually opposed, methods of logical reasoning
   a. Of a logical method; inductive reasoning SYN- to ANALYTIC *adj.*
      1705 ROBERT HOOKE *Discourse on Earthquakes* The methods of attaining this end may be two, either the Analytick, or the Synthetick. The first proceeding from the Causes to the Effects. The second from the Effects to the Causes.
   b. Of a logical method; deductive reasoning opposed to ANALYTIC *adj.* (cf. SYNTHETIC GEOMETRY *n.*)
      1833 ANON. *Edinburgh Review* Some call this mode of hunting up the essence the Analytic; others again, regarding the genus as the whole, the species and individuals as the parts, style it the Compositive, or Synthetic, or Collective.
   c. In Kantian philosophy, a statement that is true because it accords with intuition (cf. ANALYTIC *adj.*)
      1833 IMMANUEL KANT *A Critique of Pure Reason* Synthesis is that what in fact gathers the elements for cognition and unites them to [form] a certain content. Hence if we want to make a judgment about the first origin of our cognition, then we must first direct our attention to synthesis.

Each of the usages of "synthetic" I have explored so far can be traced to the early seventeenth century, when the word was etymologically rooted in Greek *syntithenai*, "put" + "together." During the Scientific Revolution, experimentalists and philosophers alike increasingly incorporated the synthetic method into analytic approaches to physics, metaphysics, and mathematics. As mathematicians and philosophers adopted empirical approaches to the natural sciences, they treated analytic and synthetic methods of proof as complementary, rather than contradictory, tools. The natural world could be grasped not only by ratiocination but also by experimentation.

Synthetic in this period also held another, wholly different sense: the "combination of parts into a whole; constructive" (*Oxford English Dictionary*). Surprisingly, "synthetic" was sometimes used to describe analytic reasoning and sometimes contrasted to it. For example, Robert Hooke writes in his treatise on earthquakes, "The methods of attaining this end may be two, either the Analytick, or the Synthetick. The first proceeding from the Causes to the Effects. The second from the Effects to the Causes."[1]

Contrariwise, an 1833 review of recent publications in logic in the *Edinburgh Review* describes two modes of Aristotelian natural classification: "Some call this mode of hunting up the essence the Analytic; others again, regarding the genus as the whole, the species and individuals as the parts, style it the Compositive, or Synthetic, or Collective."[2] Here, synthetic maps roughly onto a deductive method, and analytic onto an inductive one. In mathematics, this meaning of "synthetic" as the opposite of "analytic" is maintained in "synthetic geometry," which early nineteenth-century mathematicians viewed as a more empirical, sensory, and rigorous alternative to analytic (algebraic) geometry.[3]

Some of this semantic confusion can be blamed on Immanuel Kant. "Synthetic" first arose as a term in logic in the eighteenth century, when philosophers, Kant foremost among them, distinguished synthesis from analysis. For logicians, "the synthetic" denotes concepts "proceeding from causes or general principles to consequences or particular instances; deductive" (*Oxford English Dictionary*). Kant took the synthetic to refer to judgments that do not logically proceed from their subjects but rather extend knowledge, combining sensible intuition and thought: "synthesis is that what in fact gathers the elements for cognition and unites them to [form] a certain content. Hence if we want to make a judgment about the first origin of our cognition, then we must first direct our attention to synthesis."[4] This usage is distinguished from both empiricism and inductive thinking, which move outward from particular cases to construct general theories or principles.

This gloss on "synthetic" reflects the fact that synthetic biology is as much a field dedicated to thinking about and theorizing the ontology of life as it is an engineering project committed to rebuilding it. While Evelyn Fox Keller writes that "this classical meaning [of the synthetic/analytic division] has little to do with the ways in which the term is used today in synthetic biology,"[5] this use of "synthesis" to describe a way of reasoning does map onto a form of thinking with which synthetic biologists grapple: how might particular cases of manufactured life be extrapolated outward to demonstrate or prove general theories about what "life" is and how it might work?

Following Kant, synthetic statements cannot be true on the basis of internally constructed fixed rules or pure concepts alone but must also incorporate (synthesize) something beyond the object itself. If, for example, life is the concept under investigation, then statements about life must exceed it—perhaps in the form of sense impressions garnered from the lived world, or else intuition. In this respect, Kant aligns synthetic thinking with Newtonian physics and Euclidean geometry. Observing the chemical and physical experimentalists of the time, he noted that reason guides scientific experimentation; it is an active interrogation of nature (what historians of science term "theory-laden").

The "synthetic a priori"—nonempirical judgments that are extensive rather than internal to a logical system—has been a point of contention for much of Western philosophy.[6] What, philosophers asked, is the relation of sensation and experience to knowledge and understanding? And how might evidence gathered by the senses be taken as reliable ground for scientific judgment?

The subjects of the following chapter name themselves "DIY biologists." Like synthetic biologists, they are curious about how the living world may by refashioned to reveal some of the contours of what life is, both as a "fixed" category and as something to be intuited, impressed, and sensed through its manipulation. As such, amateur bioengineers, like their synthetic biological kin, express and embody a peculiar instantiation of the synthetic a priori (one that, I suspect, Kant would not have recognized). Rather than joining sense impressions to align intuition with categorical knowledge, the a priori is something that is itself constructed through synthesis. Synthetic biologists build particular theories into the objects of synthetic biology, which then reflect what synthetic biologists posit life has been all along, prior to their manipulations. The conceptual, the sensible, and the empirical converge, and the a posteriori masquerades as the a priori.

CHAPTER FIVE

Life Makes Itself at Home: The Rise of Biohacking as Political Action

Do-it-yourself bio: the latest mushrooming cottage industry. A computer, a credit card, and a little patience, and a person might customize a living thing. —Richard Powers, *Orfeo*, 42

Introduction: Confessions of a Biohobbyist[1]

A t this point in the book, I feel compelled to come clean. In 2003 I earned a bachelor's degree in cultural anthropology that made my mother weep for what had once seemed my bright future. At loose ends, I moved to New York City to work for technoartist Natalie Jeremijenko. There, I ghostwrote and edited articles for her *Biotech Hobbyist Magazine*. This was not a particularly lucrative venture, so I relied on the largesse of a generous girlfriend with a real job at a midtown financial company. I supplemented my meager earnings with odd hours working as a barista and bartender, alternately energizing and blunting the moods of Park Slope parents.

For years, I've been quietly yet profoundly grateful that my name was not attached to the *Biotech Hobbyist Magazine*,[2] the tone of whose inaugural issue wavers between the satirically instructive and the downright batty: "Interested in growing skin? You are not alone." This entry shilled a starting kit that customers could order online, which provided everything necessary to culture human skin cells at home. The essay ended by enjoining readers, "There are endless things to do with skin—do you want to make it glow in the dark? Do you want it to talk directly to your computer by interfacing it with silicon? Of course you do! The next project installments will explain how to splice in an amplified Great Star coral gene that will make your tissue glow cyan under UV light." A few months after I wrote this manual, I was accepted into grad school, and the next installment of the series remains to be written.

I do not make these admissions lightly, but in order to explain what about biotechnology first piqued my curiosity. That I wrote for an outlet that billed itself as "*the* place on the web for biotech thinkers, builders, experimenters, students, and others who love the intellectual challenge and stimulation of hobby biotech"[3] reveals the investments I had before ethnographically studying synthetic biology, as well as what it was about the field that first caught my attention. Namely, what arrested me at the time was that although there are many microbes that live in my home, are preserved in my fridge, and teem in my gut, it would be considered subversive, if not illegal, for me purposefully to culture microbes. The difference is not a biological one—*E. coli* might live within me, but if I want to grow it in a petri dish, I better have undergone safety training and have access to a certified BSL-1 (biosafety level 1) laboratory. Why is this so? Put otherwise, I cared then (as I do now) about what transubstantiates commonplace living things into objects upon which biotechnical control is exercised, as well as how the distinction between the two gets patrolled and managed at different moments and in various places.

My early interest also demonstrates that do-it-yourself biology ("DIYbio," as members call it), the community of amateur bioengineering tinkerers I treat in this chapter, is not the first example of the impulse to "do" biotech at home. I had the same urge years ago, as did many others, and it significantly predated synthetic biology. That being said, the current iteration of DIYbio *is* symptomatic of synthetic biology: it is made possible by synthetic biology but is nonetheless neither overdetermined by nor answerable to it. One cannot make sense of synthetic biology without also attending to its disqualified and illegitimate kin, which was already cocooned within the synthetic biology project back in 2000.

Five years after my foray into the *Biotech Hobbyist*, in the midst of ethnographically studying professional bioengineers, I once again found myself back in the domain of the biotech hobbyist. In May 2008 I was invited to a bar near the MIT campus for the first meeting of a new collective that referred to itself as "DIYbio." This was an emerging group of biotech tinkerers who were curious about how biotechnical experiments might be done at home. About two dozen other Cantabrigians were sardined shoulder to shoulder in a poorly lit back room, hoisting pints of beer and surrounded by faux-Celtic furniture and paneling. Fittingly, the bar is adjacent to a large biotech company. Some among us were graduate students and postdoctoral fellows from nearby universities, others were from biotech companies, and the rest were unattached, unaffiliated, or uncredentialed (I counted several high school students in the mix, as well as one journalist who was taking notes as frenetically as I was).

Mackenzie "Mac" Cowell, the affable enthusiast helming DIYbio, stood smiling at the center of the crowded bar. Bearded yet baby-faced, dressed in plastic-framed glasses, jeans, and a T-shirt advertising MIT's Department of Biological Engineering, Cowell blended in with a particular strain of proud hipster-nerd endemic to areas around Cambridge's Red Line of the local subway system.[4] He posed a single, deceptively simple question: "Can molecular biology and bioengineering be a hobby?" DIYbio, in the intervening years since this first meeting, has mushroomed into a network of thousands of biohobbyists. There are twenty groups in major cities in the United States and Canada, sixteen groups harbored in European cities, and a handful more can be found in other cities that have a robust technoscientific economy: Singapore, Tel Aviv, and Sydney. Members refer to themselves as DIY biologists or biohackers.[5]

The last chapter ended with biologists being jettisoned from corporate synthetic biology laboratories—biological making becoming the repetitive and mindless task of unskilled workers, while trained biologists shuffle off to middle management or do design work on their computers. In a strange twist, a community of hackers has in recent years laid claim to molecular biology not as a professional vocation but as a hobby.

In what follows, I explicate the characteristics (technical and social, as well as accidental) of DIYbio that conditioned its rise as a field that colocalizes and overlaps significantly with synthetic biology. The dividend will be a clarification of what "life" becomes when biological work gets done by nonbiologists outside circumscribed professional spaces, and when the biological project is no longer simply analytic but synthetic.

This is not the first time that the separation of design from manufacture, and of knowing from making, has spawned amateur or DIY movements, although it is the first time that such movements take molecular biology as their object. Midcentury British DIY culture, for example, followed similar lines: "Labour, in its worst-case scenario, is defined as structured, repetitive, and extrinsically rewarded. Leisure, on the other hand, is controlled and regulated by the individuals themselves, and is seen as containing intrinsic rewards. Do-it-yourself and home improvement contains all the qualities of a leisure activity whilst actually consisting of the kind of work that would be seen in another context as undesirable labor."[6]

DIYbio seeks to scale down the sort of work now conducted in large-scale biotech laboratories and to respond to industrialization, automation, and speed with craft, artisanship, and slowness. In so doing, biohackers aim to realign the politics of biological labor in order to claim rights of access to biological making and knowing. As one biohacker phrased it on the DIYbio LISTSERV, "the goal is that molecular biology should be for everyone, not just a walled garden for academic researchers in big institutions."[7]

While DIYbio is an international movement, what follows is based on fieldwork I conducted in the early months of the movement's incubation in Cambridge during 2008 and 2009, interviews conducted with biohackers, and close readings of the online DIYbio LISTSERV. Biohackers, I found, delight in troubling what it means to *do* biology, who counts as a biologist, and what constitutes biotechnical matter. They trouble the epistemic, proprietary, legal, and taxonomic contexts that shape how living stuff travels: why some biological things circulate and are understood in one way, while other, genetically and materially equivalent, living things are understood otherwise.

The Institution for the Amateur

Alongside Mac Cowell, Jason Morrison and Jason Bobe cofounded DIYbio. As an undergraduate at Davidson College in North Carolina, Cowell joined his school's International Genetically Engineered Machine (iGEM) competition team in 2005. A few years later, he moved to Cambridge, where he talked his way into a job with the Registry of Standard Biological Parts and helped to organize subsequent iGEM competitions. What riveted Cowell was iGEM's promise of democratizing the experience of biological experimentation by helping undergraduates learn how to engineer biological systems, which is work typically done only by scientists who have already completed at least a year or two of graduate school. Cowell quit working for the registry in 2008, claiming that he "wasn't learning new things" anymore, and sold his car to bankroll a new start-up community of amateur bioengineers.

Morrison studied computer science at the Rochester Institute of Technology before beginning a string of jobs as a software developer. He soon met Cowell and Bobe, and he proposed developing with them something he named "SmartLab," an interactive multitouch software technology that would automate lab protocols, annotate data, and manage multiple lab equipment at once.

Bobe had earned a bachelor's degree in molecular biology at the University of Colorado, Boulder, ten years earlier, and at the time worked for George Church as director of community outreach for his Personal Genome Project at Harvard, which was then enrolling a thousand volunteers willing to have their genomes published online, alongside medical histories and other personal data.[8] Bobe's goal for DIYbio was building a "biological weather map," which he described as a sort of "flashmob" or "Improv Everywhere" approach to tracking diseases. The idea was as follows: clusters of biohackers would converge upon an urban street, swab crosswalks, lampposts, and noses, sequence the resulting samples, and overlay the data onto an online map that would show which disease outbreaks might be brewing in major cities.

To return to DIYbio's formal inauguration in May 2008: in the brief speech that followed his astonishing question as to whether molecular biology might be a hobby, Cowell identified 1970s-era electronics hobbyists as kin, declaring that he wanted DIYbio to be "the Homebrew Computer Club of biology today." The Homebrew Computer Club (HCC) was an amateur group started in Menlo Park, California, in 1975, whose members included Apple founders Steve Jobs and Steve Wozniak.[9] Piggybacking on breakthroughs in computer programming and electrical engineering, the HCC ushered in home or personal computing (the club was not, despite the way the folktale is narrated, just young guys with big ideas tinkering in garages). The concept of personal computing, as it was forged by amateur groups like the HCC, was a direct response to the monstrously big computing machines built during World War II and the following decades. Its proponents sought to make computing an accessible, user-friendly, even domesticated technology.[10]

So too, biohackers in 2008 were reacting to biotechnology's project to mass-produce biological systems. The comparison to the HCC, which has proven astoundingly resilient, is of course pure American folklore, bearing little resemblance to reality yet nonetheless useful as a shared mythology. In this tale, a little capital and a lot of geeky enthusiasm yield technological breakthroughs and piles of cash. Instead of Silicon Valley, call this edition Carbon Valley. It orients biohackers precisely because of its hegemonic appeal, casting them as already the victors of the next big technological and commercial success story.[11] The fact that biohackers draw upon this narrative as a myth with which to make sense of their own practice—despite more obvious twentieth-century antecedents, including HAM radio operators, home chemistry kits, amateur rocketry, and even amateur breeders of experimental plants—is revealing.[12] It suggests that dormant in their social imaginary is a belief that the democratization of biotech is normatively good, and that the high road to such democratization requires bootstrapping upon the engineering ethos forged by synthetic biology.

The question that was posed at the first meeting of the HCC, as reported in the first issue of its newsletter, was "What will people do with a computer in their home?"[13] This query astonishes contemporary ears, given the ubiquity of computers in American middle-class households. Everyone in the bar in 2008 knew *exactly* what could be done with a personal computer. The question Cowell next posed to the crowd self-consciously echoed that earlier question and in so doing primed listeners to hear how absurdly self-evident it might sound in thirty years' time: "what will people do with biology in their home?" No one yet knew, but everyone had a suggestion.

The discussion that followed was lively: participants discussed how quickly they could install a public wet lab and compared prices for lab equipment—a thermocycler could be built for $10, one engineer promised, though another person worried over the cost of pipettes, which tend to retail for about $400 each.[14] One professor in attendance had a simple solution: "Steal pipettes!" Jason Kelly, who would later cofound Ginkgo, proposed that amateur biologists select a new model organism that would be well suited to amateur biology, an organism (such as moss) that would be easier to handle or see than *E. coli* or bacteriophage. Another attendee suggested yeast as a model organism because it is inexpensive, easily acquirable, and unlikely to raise the hackles of the FBI, given its long history of home cultivation for baking bread and brewing beer. Others raised possible causes for concern, such as the ambiguous legality of harboring bacterial and cell cultures outside academic and industrial laboratories. Still others asked what the endgame was: what did amateur biologists hope to accomplish? Cowell responded, "People are interested in *crafting* living things. There is a community primed for biology at home."

While the meeting might seem a bit dubious or sub rosa, venerable MIT and Harvard professors were in attendance and talked animatedly about how to get amateur biology off the ground. Their presence suggested that DIY biology is not outside synthetic biology but maintains an evolving symbiotic relationship with it.[15] It also strengthened DIYbio's newly forged and claimed connection to MIT's fabled hacker culture. Tom Knight, storied hacker of MIT's Computer Science and Artificial Intelligence Laboratory, perched precariously on a barstool beside Harvard professor Pam Silver. Together, they made suggestions veering between the practical and the mischievous; Knight went so far as to suggest that biohackers "subvert a well-respected organization like the Boy Scouts" to recruit new members.

One reason biohacking has latched on to synthetic biology's coattails relates to synthetic biology's origins in MIT's Computer Science and Artificial Intelligence Laboratory, a place legendary for its ties to hacking. MIT is sometimes credited as the fountainhead of hacking; the term "hack" derives from MIT undergraduate pranks.[16] Such stunts have nothing to do with computing, although "hack" has since been applied to computer hacking. Rather, at MIT, hacks mean playful, clever, anonymous, and (this should not be underestimated) *funny* engineering feats or stunts.[17] Though biohacking makes alliances with computer hacking, it also shares valences with this sort of crafty but innocuous play.

Cowell openly models DIY biology on hacker practices, ending his public presentations with the slogan "real hackers write DNA," a sentiment echoed by the synthetic biologists who founded Ginkgo Bioworks— the business cards they distribute label them "DNA hackers" and their phone number is HACK-DNA. At the 2007 MIT commencement, President Emeritus Charles Vest lauded biohacking in his speech, announcing, "Already the term 'biohacking' is heard along the Infinite Corridor [an indoor hallway connecting the east and west sides of the MIT campus]. 'Biohacking' . . . just think about the significance of that term! It of course heralds the advent of synthetic biology, the fusion of engineering and biology to design and build novel biological functions and systems."[18] Vest confuses synthetic biology and DIY biology under the shared tag of "biohacking," demonstrating the extent to which both synthetic biology and DIY biology draw heavily upon the fantasy of the hacker, whose mythical origin and playground remain the MIT campus.

Following the first meeting, DIYbio quickly arranged itself online as a congregation of local communities of amateurs, first in Cambridge but soon elsewhere. Cowell and Bobe explained the purpose of DIYbio as follows: "We want to be the institution for the amateur, to provide access to the resources that you can't get if you're not in one of the traditional institutions like academia or industry: access to experts, like peers who can tell you how to get around that little problem. Or the literature, which unfortunately isn't free. Gentle oversight."[19] Nearly everyone participating in the Cambridge branch of DIYbio at the time had or was pursuing an undergraduate or graduate degree in a scientific field, a fact that complicates what counts as being an "amateur."

Biohackers locate themselves within twenty-first-century "maker culture," yet practitioners also narrate the prehistory of their field using computer hacking as a prior example and relevant model for "hacking" biology. DIYbio is adamantly not citizen science, yet it is not pseudoscience either. Citizen science is the sort of work nonprofessionals do, often at the behest of large academic or national laboratories seeking to gather masses of data. Citizen science is, most often, observatory labor: amateur astronomers identifying stars and other celestial events, or interested amateurs allowing their computer, during its idle time, to crunch data collected by SETI to sort intergalactic signals from background noise.[20] The affect of such work is participatory, populist, and enthusiastic. In contrast, though democratic, DIYbio is noticeably antagonistic, roguish, and mischievous in tone.

Pseudoscience, as historian of science Michael Gordin has documented, is another thing entirely. Practiced by people with questionable credentials, it is hard to pin down except by the term "pseudoscience" itself, which is most often hurled as an epithet by scientists when they feel their authority threatened.[21] DIYbio is not pseudoscience, either—biohackers rarely forward "crackpot" theories that oppose biological knowledge, and there is no need to test whether biohackers' theories are falsifiable. Their truck is not with Popper.[22] Differentiating DIY biology from citizen science and pseudoscience (elsewhere referred to as "outsider" science, as an analog to "outsider," or folk, art)[23] is necessary because, as opposed to these other permutations, the goal of DIYbio is for amateurs or nonprofessionals to make, not new *theories*, but new *things*.[24] In so doing, they operate against hegemonic biotech and the legal and proprietary corralling of biological media and labor.

In this respect, DIYbio, like the late nineteenth-century Arts and Crafts movement, seeks to regain lived experience, a romantic attachment to nature, and artisanal proficiency in the wake of mass industrialization: it is "a response to the tensions and ambiguities that were spawned as mass production both created the role of the designer and also removed designers from direct contact with making. . . . [It] encourage[s] amateur practice."[25] In a latter-day response to biological mass production, biohackers seek direct contact with designing and making living substance and reclaim biological making as an amateur practice not limited to professional synthetic biologists.

Attending to biohackers means tracking a biology not practiced by biologists, not performed in laboratories, nor exercised upon living materials typically defined as "biotechnical." Taking seriously amateur groups like DIYbio (and their various historical precursors in physics, chemistry, astronomy, and biology) is one way science studies scholars can redefine the contours of scientific practice independent of the boundary work professional scientists do when policing the frontiers of legitimate science.[26] Instead of taking as given what is *inside* and *outside* science, science studies might appraise new forms of public participation in science that demonstrate gradations of scientific working and thinking.[27]

We, the Biopunks

"I need help on isolating *Vibrio fischeri* from squid and fish." "Dangers eating fluorescent bread?" "Seeking a dye for diffusion in organs." "In Vitro Tuna." "How would you feel about genetically engineering humans?" Such subject lines are culled from the Google group LISTSERV, started in April 2008, with which the DIYbio group communicates.

Only a few dozen of the over 3,550 members post regularly to the list.[28] It is a forum in which people working on home biological experimentation may seek advice when they run into problems, share successful protocols, post links to articles that they think will be of interest to other biohackers, and argue. Topics of debate include how a hobbyist community can regulate best practices and lab safety among its members, proper list etiquette, and how members of the group should present themselves and formalize DIYbio's aims. Biohackers want to master the technical laboratory skills inculcated in apprenticeship-based lab pedagogies:[29] isolating DNA, doing PCR (polymerase chain reaction), running gel electrophoresis, culturing bacteria and cell lines, and genetically modifying organisms.

Many of the conversations on this list also focus on developing the tools needed to bioengineer at home, where to find cheap lab equipment, and how to build inexpensive versions of professional lab equipment. Indeed, Len Sassaman, a well-respected coder and biohacker who maintained an energetic presence on the DIYbio list in its first two years, even equated biohacking with the work of crafting lab equipment from readily available materials: "*The* goal of DIYbio, for me, is to reduce as much as possible the specialized equipment handicap for those who choose not to take the degree track/academic institution approach. . . . DIYbio is a hardware hacking endeavor at its core, and it's the hardware hackers working hand-in-hand with the protocol authors who are laying the groundwork for making this a field open to anyone with the drive to become great at it."[30]

Members offer one another advice on where to buy agar, a polysaccharide gelling agent commonly used as a medium for culturing microorganisms (you can find it in most Asian and Indian groceries). They suggest to one another readily available alternatives to lysogeny broth, a liquid medium for growing bacteria (chicken broth being one adequate substitute). They also share information on the best outlets for cheap lab equipment. One list member received attention and accolades after he designed an inexpensive attachment that turns a Dremel rotary tool into a functioning centrifuge capable of spinning over 33,000 rpm.[31]

Scouring eBay, biohackers bid on gel boxes used to perform gel electrophoresis, shaker tables, vortexers, pipettes, and centrifuges at steeply reduced prices ($500 gel boxes, for example, can be bought online for $9.95). Some prowl biotech companies that have gone bankrupt: in 2009 Cambridge biohackers pillaged Cambridge's defunct Codon Devices, "watching their dumpsters" for discarded lab equipment.[32] Those more interested in tinkering cobble together equipment from odds and ends—assembling an incubator from a plastic cooler, a thermostat, and a light bulb; repurposing a terrarium as a thermocycler, and an old turntable as an orbital shaker. Flip around the lens on a webcam, and you have a $10 microscope.[33] They proudly share photographs of the labs they have improvised in home garages. Making hay of the financial crisis, biohackers work athwart Big Science, operating outside it while poaching its resources and gadgetry, such as lab supplies and equipment.

Biohackers are keenly aware that their kitchen and garage laboratories, after 9/11, anthrax scares, and the PATRIOT Act, could raise concerns about both biological terrorism and biosafety, as the personae of the biohacker and the bioterrorist get confused in the eyes of various publics (including amateurs themselves).[34] In 2011 biohackers attempted to draft a brief "code of ethics" stipulating that they "emphasize transparency," "adopt safe practices," and "respect humans and all living systems."[35] While the field has not sufficiently advanced to pose meaningful threats to the public, dangers cannot be dismissed outright. Some synthetic biologists propose technical solutions to this risk, such as requiring DNA synthesis companies to screen orders for potentially dangerous genetic material, while policy analysts worry that the long arm of the government may soon reach into academic peer review when reported results are potentially "dual purpose." This was the case in 2005 after the sequence of the 1918 flu was "resurrected" and published in *Nature*.[36]

Anthropologist Paul Rabinow notes that technical solutions cannot adequately address social and ethical problems, and that thinkers on both sides of the conversation about ethics in synthetic biology (those who support either the precautionary principle or unfettered bioengineering) are divided by polemics. He and his coauthors instead espouse a "vigilant pragmatism" that consists of "rigorous, sustained and mature analysis of, and preparation for, the range of dangers and risks catalyzed by synthetic biology and DIYbio."[37] Biohackers and ethicists alike suggest that because synthetic biology and DIYbio are open-ended and uncertain fields, they require more than "common, deontological or utilitarian calls for anticipatory policy, risk assessment, code[s] of ethics or legal and ethical prudence."[38] Rather, the field's safe and ethical practice might instead be regulated *experimentally*, such that policy operates alongside and keeps pace with technological change.[39] Hype and worry are polarized yet compatible affective stances—both look forward to imagine technologies as they may be rather than concentrate on what they now are.

Science studies scholars who have written about amateur science focus on how nonscientists gain competence, proficiency, or credibility in discussing and deploying scientific *facts*, and they characterize scientifically engaged amateurs as "lay-experts"[40] or "amateur-experts."[41] In appraising "orthodox" science, however, anthropologists and historians now overwhelmingly attend to practice—what scientists *do*. There is a disjuncture here. Scholars examining how science is received by nonscientists still privilege the construction of semiscientific personae through the deployment of scientific "facts" rather than study performative communities of practice. But biohackers are not after facts so much as they are after acts—"doing" hobbyist biology, on their view, means building biological things and making do with discarded or repurposed equipment, not demonstrating facility with biological concepts.

Meredith Patterson, one of the most active contributors to the DIYbio LISTSERV and among the highest-profile biohackers (both among other biohackers and in the popular media), is best known for two projects: first, she inserted green fluorescent protein into yogurt bacteria to make fluorescent yogurt, and second, she worked to assemble a related system in which the presence of melamine triggered the expression of green fluorescent protein as a handy and inexpensive indicator of melamine-tainted food.[42] Patterson, a computer programmer and sometime science fiction author, drafted "A Biopunk Manifesto" expressing why she chooses to experiment with bioengineering at home, which she read aloud at a UCLA symposium in 2010. In it, she declares:

> We [the biopunks] reject the popular perception that science is only done in million-dollar university, government, or corporate labs; we assert the right of freedom of inquiry, to do research and pursue understanding under one's own direction, is as fundamental a right as that of free speech or freedom of religion. We have no quarrel with Big Science; we merely recall that Small Science has always been just as critical to the development of the body of human knowledge, and we refuse to see it extinguished. . . . A thirteen-year-old kid in South Central Los Angeles has just as much of a right to investigate the world as does a university professor. If thermocyclers are too expensive to give one to every interested person, then we'll design cheaper ones and teach people how to build them.[43]

Note Patterson's repeated invocation of "freedom," "free speech," and investigation as "rights." Though freedom is capacious enough to mean many things, she here describes freedom of inquiry as a form of self-expression. Such a stance reflects, as Gabriella Coleman puts it, "a long-standing tension within liberal legal rights but also offer[s] a targeted critique of the neoliberal drive to make property out of almost anything." Furthermore, Patterson's manifesto (like the philosophy of Free/Libre and Open Source Software) critiques liberalism from within a liberal framework "by asserting a strong conception of productive freedom in the face of intellectual property restrictions."[44]

To put a finer point on Patterson's biopunk ethos, DIYbio is a response to contemporary biotechnology in which *practitioners perform an embedded critique of the political economy of the life sciences.*[45] In what anthropologist Chris Kelty has termed a "recursive public," they mobilize around producing the technical conditions of their own existence. One Cambridge biohacker in her midtwenties made this stance explicit when she said to me of DIYbio, "It's not some black magic guarded by white-coated minions at Monsanto that steal your money and your soul." In this sense, the products of DIYbio—the stuff amateur biologists make, like yogurt bacteria bearing the gene for green fluorescent protein—are beside the point. It is the *act* of making such things, in unregulated spaces, by self-trained and uncredentialed amateurs, that is the means by which biohackers leverage synthetic biology to critique industrial biotechnology and to claim both access to and ownership of biological substance.

Biohackers claim to be trying to "democratize" biology: they discuss "trying to democratize biotech," "democratizing science," and "democratiz[ing] the tools of creation."[46] Such an optimistic, liberal, and liberatory reading is insufficient, however, because what they are doing is more critical and interesting: their work is premised on the claim that the biological is not something cordoned off in labs but is instead quotidian, personal, and apprehensible. This is amateur biotechnology as a mode of political action, in which practitioners frame doing biological research as a right rather than a privilege conferred with a PhD. I call this practice "political action" because it engages and grapples with issues of rights of access (to biological substances, tools, and techniques), legitimacy (who is a biologist?), and public participation (what does it mean to be a member of science's "public"?), and with moral questions (who should do biology?). Biohackers address such questions to (and against) the reigning academic-industrial-governmental complex of life sciences research, including academic synthetic biology labs, private laboratories, and governmental organizations that regulate what it means to do biology safely and lawfully.

Science as Vacation[47]

To understand the roots of biohackers' critique of biotechnology requires circling back, before the beginning of DIYbio, to the origins of synthetic biology. The amateurization of bioengineering in reaction to biotechnology was written into synthetic biology's charter before the field had even gained a foothold as a scientific community. In 2000, before Endy moved to MIT, he and Rob Carlson worked back-to-back at lab benches at the Molecular Sciences Institute, a Berkeley laboratory founded by Sydney Brenner. While putting in long hours at the lab attaching DNA tails to proteins and sequencing the genome of pufferfish, Endy and Carlson had plenty of time to chat and commiserate. One outcome of their late-night conversations was a letter they coauthored (with Roger Brent, then the president of the institute) and mailed to the Defense Advanced Research Projects Agency.[48] The letter joins ideas about the ownership of biological materials and of information to projections about who can and should build new biological systems. The authors predict, following the Open Source software movement, that "the Open Source Biology community will rely on individuals and small groups of people to take charge of (and receive credit for) maintaining and improving the common technology, open to all, usable by all, modifiable by all." Although currently, the authors note, bio-technology "researchers [who] possess and can work with these [genomic] components to achieve desired ends are confined to high-end academic labs and corporations," that state of affairs will not last long.

As an example of trends in the opposite direction, they cite a *Scientific American* Amateur Scientist column that taught readers how to perform PCR and gel electrophoresis on DNA samples. The authors next describe recent developments in bioengineering, many of which were made possible by the sort of radical deskilling of laboratory work overtaking much industrial biotechnology and bioengineering, extrapolating from that trend to the near future. They point out that while molecular biology protocols were once complicated enough to require that practitioners have PhDs, the growing market and use of DNA kits containing color-coded reagents now allows the same protocols to be completed by undergraduates. Alongside such ready-made kits, the increasing speed and lower cost of DNA synthesis meant, they predict, that soon DNA synthesis "will move from academic labs and large companies to smaller labs and businesses, perhaps even ultimately to the home garage and kitchen."[49]

Carlson, Brent, and Endy posed this transition to domestic and amateur biology as a direct response to the behemoths of biotechnology: "In plant biology . . . there are really only four corporations that control collections of reagents and patent rights general enough to allow construction of most desired transgenic plants (Monsanto, Aventis, Novartis, and DuPont/Pioneer Hi-bred). For example, DuPont owns the rights to the most widely used site specific recombination system, and other companies need to spend millions of dollars working around the existing patents, or forgo the advantages that site specific recombination brings."[50] Carlson, Brent, and Endy hoped, when they drafted this letter, for a future in which sped-up and inexpensive DNA synthesis technology would decentralize and democratize bioengineering so that it could operate outside the intellectual property regimes that reign in biotechnology and pharmaceuticals. This same hope was still apparent in 2008, when biohackers would also begin narrating their aims as operating against the way intellectual property in biotech forecloses access to biological materials and experimental techniques.

In a later article, Rob Carlson would further forecast: "Biological engineering will proceed from profession, to vocation, to avocation, because the availability of inexpensive, quality DNA sequencing and synthesis equipment will allow participation by anyone who wants to learn the details. In 2050, following the fine tradition of hacking automobiles and computers, garage biology hacking will be well under way."[51] While the prediction that soon bioengineering will be the domain of garage hackers spoke to the democratization of biotechnology, embedded in Carlson's language is also the attendant hope that bioengineering, rather than being a skilled job, will become a hobby, with the pleasure and delight that such leisure activities entail.

Experiments in the First Person

A month after the first DIY biology meeting, on a Thursday evening, I arrived at Beta House, a communal workspace for hackers and software developers in Cambridge. This would be the first time DIYbio members would collaboratively try their hands at domestic bioengineering. That night, the aim was a "DNA Extraction in the Kitchen," following a protocol published in *Make Magazine*, a quarterly about DIY and amateur technology projects. The author played up the dramatic elements of DNA isolation, writing: "The properties of this massive molecule are so mysterious and wondrous that most folks assume only the enlightened priesthood of laboratory biologists can extract and study it. Not so. In fact, anyone can extract, purify, and experiment with DNA at home."[52]

A handful of would-be biohackers, many of whom I recognized from the first meeting or from previous encounters at synthetic biology events, drifted in and out of Beta House over the course of the night. Some participated in the DNA extraction protocol; others enjoyed the free beer and conversation. Among them were undergraduate and graduate students from the Boston area, recent graduates in the midst of founding start-up companies, and one transhumanist.[53]

Before beginning that night's experiment, we reviewed the DNA isolation protocol while seated around a central island that separated the kitchen from the clusters of desks and couches crowding the work area. Cowell drew a flowchart of the protocol on a whiteboard, we gathered together the necessary materials, and then we set to work. We mixed a buffer solution of distilled water, salt, baking soda, and dish soap, which is used to burst open cells, releasing the DNA into solution. As a source of DNA, some people had brought fresh fruit, chopping up and grinding apples into a puree. Others ground up oatmeal, and I worked with a young software designer studying at a nearby university to extract human DNA: we both chewed lightly on the inside of our cheeks before spitting into a red plastic cup.

Once everyone had minced, pulverized, or chewed their sources of DNA into viscous solutions, we added to the buffer a few drops of contact lens solution, which would help break up the proteins that would burst from the cells with the DNA. We put this solution on ice so that when we added the DNA samples to it, the DNA wouldn't degrade too quickly. One attendee had stolen a handful of plastic Corning test tubes from a Harvard lab, and we loaded three tubes with apple, oatmeal, and human samples. We then added two teaspoons of buffer to each tube, shaking to mix. The next step was to separate the liquid in the sample from the solid matter. In lieu of a centrifuge, we used paper coffee filters. The final step was to precipitate the DNA out of the buffer solution, which requires lowering the salt concentration using alcohol. We pipetted isopropyl alcohol out of its bottle with a plastic drinking straw and dribbled it along the inside edge of each test tube so that it settled in a layer above the buffer. At the interface between the buffer and the alcohol, the DNA precipitated out of solution, as a cloudy tangle of white filaments floating in the rubbing alcohol (fig. 5.1).

When the DNA became visible, people took out iPhones and digital cameras to photograph one another staring into test tubes held up to the light, aping the scientist's headshot so common in newspapers and glossy magazines. Everyone around me was grinning, excited; some laughed in undisguised wonderment. Cowell and the others had planned to start working immediately on gel electrophoresis of our isolated DNA, but the protocol took longer than anticipated, so electrophoresis was postponed until the next meeting.[54] We dipped toothpicks taped to the ends of straws into our test tubes, and the suspended DNA clung snottily to the end of each toothpick. We then dropped the toothpick, with its attached DNA globule, into a second tube of isopropyl alcohol, which could chill indefinitely in the freezer.

What was manifested in these test tubes when a fibrous genetic gloop coagulated out of solution? What sort of DNA had materialized: was it the rhetorical "magnetic tape" or "code" containing the "secret of life"?[55] The linguistic, cryptographic, or mystical "Word" or "Book of Life"?[56] An organismic "blueprint,"[57] a "code-script,"[58] an "artificial intelligence,"[59] or

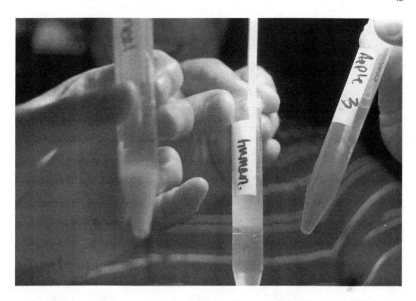

FIGURE 5.1. Isolated samples of apple, oatmeal, and human DNA, extracted during the second DIY biology meeting, June 2008. The human DNA is the author's own. Photo by Mackenzie Cowell. Licensed under a CC-A 2.0. Full terms at https://creativecommons.org /licenses/by/2.0/.

"machine language program"?[60] Was it DNA as information, map, "code of codes," "master molecule," or any of the other hoary metaphors with which nucleic acids have been freighted?

Certainly, many of the biohackers there that night had been reared in such rhetorics; they are ubiquitous in the life sciences.[61] But that does not explain why they were laughing. Instead, this sounded more like the laughter of recognition, seeing themselves in the physical molecule they had isolated and delighting in the deliberative and interactive "first-person experimentation" of amateur biotechnology.[62] They took pleasure in the creative and artful way they had distilled their own genomic selfhood using everyday household items. It was therefore a gleeful laying claim to their genomically grounded identities, notably a genome that was *not* dematerialized or sequenced but was instead a substantive molecule visible in solution.[63] Their laughter also reverberated from a disconcertingly reflexive subjecthood, predicated upon what anthropologists and sociologists of biology alternately have called the "genetic self,"[64] "biosociality,"[65] "biotechnological individualism,"[66] and "genetic citizenship,"[67] in which identity, selfhood, and relatedness are conflated with genetic material.

Conclusion: Crafting Life

Biohackers operate using the technologies, investments, markets, and stances that drive synthetic biology—foremost among them the intercalation of making and knowing. Recall that at Cambridge DIYbio's first meeting, Cowell articulated biohackers' investments by announcing, "People are interested in *crafting* living things." Indeed, one biohacker told me during a meet-up at MIT in 2008 that biohackers are riding "the coattails of baking bread in the kitchen, making sauerkraut, cutting flowers, and brewing beer. It's on that continuum." Another I spoke to, who enjoyed both cooking and biology, stated bluntly that, for him, "biology is a kitchen science." And Bobe compared the project of amateur biology to urban farming—a sort of intensely local, back-to-the-land variety of maker culture that seeks to cultivate sustainability and minimize dependence on large agriculture conglomerates.

Such statements align biology with more domestic varieties of DIY or maker culture and the implicit care, attention, and tending to life that such crafts entail. Before the last decade, such activities might have been classed (and devalued) as crafty self-sufficiency or even more simply as feminine labor and domesticity: bakers who maintain and care for yeast strains as sourdough starter, amateur taxonomists foraging for mushrooms and seaweed, backyard beekeepers, and those among us who carefully calibrate pH and temperature when brewing beer, infusing bitters, or fermenting *kombucha* or *kimchi*.

Taking a crafty approach to biology, these biohackers perform biological experiments as an artisanal process in which epistemic knowledge is generated through practical proficiency. Pamela Smith has described the relation of artisanship to scientific knowledge during the Scientific Revolution as "'artisanal literacy,' which had to do with gaining knowledge neither through reading nor writing but rather through a process of experience and labor."[68] In this respect, biohacking is one more rejoinder to the entanglement of making and knowing (*techne* and *episteme*, or *practica* and *theoria*) that animates synthetic biology. It is, biohackers claim, on a continuum with other domestic crafts in which practical skill (*techne*, or knowing-how) requires dexterity, virtuosity, and flexible ad hoc contingent art.

Yet, as both anthropologists of craft and laboratory ethnographers remind us, embedded in such practical and embodied skills—and inculcated through them—is another sort of knowledge (*episteme*, or knowing-why).[69] Artisanal skill contends with and reacts to mass production and automation, sometimes promising a smaller, local, or "more authentic" rejoinder to industrialized society (think of hipsters with backyard chicken coops). This is the case with biohacking, which demands a "Small Science" or a "Slow Science"[70] to offset the scaling up and acceleration of biotechnology, bioengineering, synthetic biology, and allied sciences. Yet craft is also the domain of knowledge hard-won through the slow accretive process of learning how to perform an embodied task. This is the promise that lures biohackers: that learning *how* to conduct a biotechnical protocol will offer them a better feel for and understanding—a literal and cognitive incorporation—of the living substances that compose and surround us.

Life Embryonic and Prophetic

syn·thet·ic (sĭn-thĕt'ĭk)
adj.
(In most senses opposed to EMBRYONIC and PROPHETIC *adj.*)

6. *Biol.* Multiple senses.
   a. Pertaining to the theory by which primitive organisms exhibit in combination characteristics that appear singularly in later types.

> 1849 LOUIS AGASSIZ *Twelve Lectures on Comparative Embryology* Synthetic types are those which combine in a well balanced measure features of several types occurring as distinct, only at a later time. Sauroid Fishes and Ichthyosauri are more distinctly synthetic than prophetic types.

   b. Compound. Of a biological subfield: as distinct from *analytic* biology, or pertaining to the manufacture of living systems, esp. from inorganic substances (cf. SYN- *recombinant DNA* or *Artificial Life*) (*Obs. Rare.*)

> 1912 STÉPHANE LEDUC *La Biologie Synthétique* La biologie synthétique représente une méthode nouvelle, légitime, scientifique; la synthèse appliquée à la biologie est une méthode féconde, inspiratrice de recherches; le programme consistant à chercher à reproduire, en dehors des êtres vivants, chacun des phénomènes de la vie suggère immédiatement un nombre infini d'expériences, c'est une direction pour l'activité.

> 1976 ROBERT SINSHEIMER "Recombinant DNA—On Our Own." With the advent of synthetic biology we leave the security of that web of natural evolution that, blindly and strangely, bore us and all of our fellow creatures.

> 1994 ANON. (SANTA FE INSTITUTE) [Artificial Life] complements the analytic approach of traditional biology with a synthetic approach in which, rather than studying biological phenomena by taking apart living organisms to see how they work, researchers attempt to put together systems that behave like living organisms. Artificial life amounts to the practice of "synthetic biology."

Naturalist Louis Agassiz, founder of Harvard's Museum of Comparative Zoology, enjoyed a colorful career spanning Brazilian natural history expeditions, a lively correspondence with P. T. Barnum, and a pet bear accidentally let loose to rampage up Massachusetts Avenue, much to the vexation of Cantabrigians. He parsed the living world into three types of organisms: "embryonic," "synthetic," and "prophetic" types. By "synthetic," he meant something different than what it now means. "Synthetic," he explained, defined those organisms "which combine in a well balanced measure features of several types occurring as distinct, only at a later time," including among such organisms the sauroid fish and ichthyosaurs.[1] By this definition, the bodies of synthetic organisms exhibit structures that will later be differentiated in future "higher" organisms but that are combined in their original forms. He used the existence of such "synthetic" and "prophetic" types as evidence with which to argue against theories of natural selection—including Darwinian theory—by positing that all biological diversity exists in proto-states in early life-forms. This form of "synthetic" simultaneously points toward organisms lost to natural and human history and those yet to come.

The first application of "synthetic" to name a subdiscipline of biology appeared only in 1912, the coinage of biophysicist Stéphane Leduc in his *La biologie synthétique*. Leduc performed a series of experiments in which he simulated living processes and lifelike forms in abiotic matter, growing crystal salts and percolating ink droplets in water. Evelyn Fox Keller writes, "Leduc's aim was to demonstrate the continuity between living and nonliving matter, and to do so incrementally. Obviously, these were not living organisms. Yet they did bridge the gap with living systems in one respect: They looked like organisms. That is, Leduc had demonstrated that structures that resembled the forms of living organisms—at least in their outward appearance—*could* arise spontaneously, from brute matter, without help from either a designing deity or a vital force."[2] Keller points out that Leduc's use of "synthetic," as well as that of other biologists working in the early twentieth century, was inherently ambiguous, "meaning both constructed and artificial."[3] Leduc obscured the distinction between organic and inorganic by producing synthetic living forms in abiotic media. Today's synthetic biologists, however, do the reverse, working to make biotic media conform to design principles culled from electrical, chemical, and computer engineering. In so doing, they question divides not only between organic and inorganic but also between natural and artificial, found and made.

Eighty years after Leduc, "synthetic" would again be appended to "biology," this time to describe lifelike forms rendered in silicon rather than salt. A 1994 Santa Fe Institute mission statement announces that Artificial Life "complements the analytic approach of traditional biology with a synthetic approach in which, rather than studying biological phenomena by taking apart living organisms to see how they work, researchers attempt to put together systems that behave like living organisms. Artificial life amounts to the practice of 'synthetic biology.'"[4] Distinguishing between an analytic and a synthetic approach to biology, Tom Ray, the developer of the Tierra program, which simulated evolutionary algorithms, defined Artificial Life as "the enterprise of understanding biology by constructing biological phenomena out of artificial components, rather than breaking life forms down into their component parts. It is the synthetic rather than the reductionist approach."[5] "Synthetic" is again ambiguous: it refers both to the process of "constructing" as a method "of understanding biology" (as opposed to the analytic) and to a form of life that is "artificial" rather than "natural." This equivocality inheres in contemporary synthetic biology.

"Synthetic" bears one last meaning in the life sciences. Molecular biology and genetics in the 1970s and 1980s developed two technologies, both of which were believed to be hastening an age of "synthetic biology." The first of these technologies was recombinant DNA technology, which allowed researchers to combine genes from multiple organisms or species into a single genome. In 1976 Robert Sinsheimer, a biologist then at the California Institute of Technology, wrote of the brand-new technology: "With the advent of synthetic biology we leave the security of that web of natural evolution that, blindly and strangely, bore us and all of our fellow creatures. With each step we will be increasingly on our own."[6] Sheldon Krimsky, a bioethicist who closely followed political debates surrounding the use of recombinant DNA technology in the 1970s, titled a paper about recombinant DNA "Social Responsibility in an Age of Synthetic Biology"; a 1979 *Nature* review of a book about recombinant DNA policy debates ends: "the recombinant DNA debate has conveyed to a wide audience the vast potential of the 'synthetic biology' opened up by recombinant DNA techniques."[7]

The second technology to which I refer, DNA synthesis, allows researchers to use machines to assemble new genetic sequences not found in organisms—to put together genetic material one nucleotide at a time. Synthetic biologists now identify DNA synthesis as a technique by which they may shuttle between digital information (electronic sequences) and biotic matter (living molecules). Recombinant DNA and DNA synthesis are together the paired fundamental enabling technologies of current synthetic biology, allowing researchers to stitch together coding regions from disparate organisms and to synthesize new genetic material. Echoing the warnings of Sinsheimer and Krimsky, synthetic biologists imagine a future in which DNA synthesis allows them to depart from the "security" of natural evolution. As they pluck extinct organisms from frozen bogs and jars of preservative, they hope to resurrect species long gone. This fantasy condenses the three living forms nominated by Agassiz in 1848: life-forms embryonic and prophetic are now joined in the synthetic.

Latter-Day Lazarus: Biological Salvage and Species Revival

The desire for the presence of the most ancient is a hope that animal creation might survive the wrong that man has done it, if not man himself, and give rise to a better species, one that finally makes a success of life.... The rationalization of culture, in opening its doors to nature, thereby completely absorbs it, and eliminates with difference the principle of culture, the possibility of reconciliation. —Theodor Adorno, *Minima Moralia*, 115–16

From miniature fragment comes total reanimation of the extinct creature; the beauty of miniature is destroyed for the creation of the gigantic theater of the sublime. —Svetlana Boym, *The Future of Nostalgia*, 35

Back from Extinction

Shirley Temple Three, a Bread Island Dwarf Mammoth, "is waist high, with a pelt of dirty-blond fur that hangs in tangled draggles to the dirt. Its tusks, white and pristine, curve out and up. The forehead is high and knobby and covered in a darker fur."[1] While Dwarf Mammoths went extinct ten thousand years ago, Shirley Temple has been resurrected by *Back from Extinction*, a popular cable television show. The show's host, rather than killing Shirley, as most creatures are usually dispatched after an episode airs, hides her in his mother's home in rural Georgia, where the creature languishes in the summer heat, fed mixed nuts. "Yanked out of its own time," Shirley's hair begins to fall out. She refuses food and her gray hide scabs over. One night, her caretaker leaves the backdoor open. By morning, Shirley has escaped, turned loose into the arid Georgian backwoods.

This scenario was concocted by writer Thomas Pierce and appeared as a short story in the *New Yorker* in 2012, yet it fictionalizes a scientific promise already made by Harvard synthetic biologist George Church, who hopes to use synthesis technology to "resurrect" a woolly mammoth.[2] Regular scientific reporting on ancient DNA genomics research feeds Pierce's preoccupation with mammoths. He commented to a *New Yorker* reporter, "For all I know there's a team of scientists out there right now who are on the verge of a major announcement about their cloned mammoth. Well, I doubt anyone is that close, but it's conceivable. That aspect of the story—the cloning—didn't seem too outlandish."[3] While Pierce is right that no research team is on the cusp of cloning a mammoth, synthetic biologists—foremost among them George Church—are now teaming up with conservation biologists and ecologists in hopes of doing just that: reviving passenger pigeons, woolly mammoths, and Neanderthals.

This chapter is a companion to chapter 1. While this book began by focusing on synthetic biologists who streamline bacteriophage genomes to render them more comprehensible, thereby making "new" life-forms, I now turn to their attempts to resuscitate old life-forms. If the project of synthetic biology is to make new life-forms in order to better understand how life works, where does de-extinction fit into that project? What is to be gained from reviving creatures that have already lived and died? If synthetic biologists seek to make "better" forms of life, to what end would they resurrect those species that one might say have already "failed" at the evolutionary game, disappearing from the planet as victims of changing climactic conditions or predation?

The De-extinction Project, as it is called, is metonymic of synthetic biology's overarching agenda, projecting in science-fictional extremes the more orthodox themes of synthetic biology that sound throughout this book: design and creation, purity and hybridity, value and promise. I here reexamine the ways in which synthetic biologists rework and redefine notions such as species, kind, novelty, and life in their efforts to build living things. This chapter is also an outré glimpse into what happens when the logics, rhetoric, and tools of synthetic biology are taken to fantastical limits and become public provocations.

I first learned about the De-extinction Project while speaking with Church in his Harvard Medical School office. Church is a larger-than-life figure in synthetic biology and genomics, and his frame is similarly imposing—towering, broad-shouldered, with a shock of white hair and a full beard. He speaks articulately and earnestly, the regular cadence of his voice broken by occasional bouts of hearty laughter. For over forty years, Church has had a nearly uncanny premonition about "the next big thing" in genomic technology, which he then pursues avidly. As a graduate student studying with Walter Gilbert at Harvard beginning in 1977, he developed the first genetic-sequencing technology and was among the first scientists to plot what would become the Human Genome Project.

Since then, Church's lab has developed new methods for accelerating DNA sequencing and synthesis, most recently setting its sights on the CRISPR/Cas9 system. He has attracted popular attention for several of his provocative and mediagenic genomic sequencing initiatives. He employs high-throughput genomic analysis and synthesis in projects such as the Personal Genome Project, which is currently sequencing 100,000 human genomes and posting them freely online alongside individuals' medical histories and records. He is also an avid self-tracker, monitoring his health, weight, and nutrition (he has been an outspoken vegan for many years), all of which he makes available to the public on his website.

His current plans to resurrect a woolly mammoth are merely the most extreme of a series of efforts that conservation biologists, ecologists, and zoologists have made in the last decade to "clone" endangered wildlife.[4] Artificial reproductive technologies have enjoyed mild success in generating chimeras from endangered and rare species. An endangered Southeast Asian bovine named the banteng was born using these techniques in April 2003 and is now exhibited in the San Diego Zoo. Researchers at the Audubon Center for Research of Endangered Species (ACRES) in New Orleans successfully produced three litters of endangered African wildcats between 2003 and 2004.[5] Such technologies build on long histories of cryopreservation of endangered species, which reach back to the 1960s, when Kurt Benirschke began freezing the somatic cells of wildlife.[6]

None of these projects is reliant on synthetic biology; indeed, they are demonstrative of the long-standing relation of biotechnology to agriculture and animal husbandry enterprises. Neither are technologies such as somatic cell nuclear transfer (SCNT) limited to endangered wildlife. In many cases reproductive techniques that were first developed in agriculture and zoology have been appropriated for human reproductive technologies such as artificial insemination, gestational surrogacy, embryo transfer, and preimplantation genetic diagnosis: they have been exported from zoos to fertility clinics. Breeders and ranchers used frozen semen for artificially inseminating livestock decades before cryobiology became a key tool for storing gametes and embryos for human reproductive technologies.

In recent years, however, scientists have set their sights on producing chimeras from not only endangered species but also extinct ones. Synthetic biology becomes relevant to this story in the context of de-extinction efforts in which whole frozen cells or nuclei cannot be recovered from live animals or animal carcasses, which are too old or too deteriorated. In such cases, synthetic biologists can offer their services sequencing small snippets of degraded DNA. They then plan to use multiplex automated genome engineering (MAGE), a DNA synthesis technology that can make new fragments of old genetic material and stitch them back together. They hope next to port chunks of DNA from endangered or extinct species into the whole genomes of living close relatives, which would yield biological chimeras.

References to Michael Crichton's *Jurassic Park*, the science-fictional account of geneticists reviving dinosaurs and turning a nature preserve into a tourist attraction, are understandably common in scientists' and popular portrayals of de-extinction, even as the same scientists insist that reviving dinosaurs is impossible given the rate at which DNA degrades. Whereas the plot of *Jurassic Park* hinged on dinosaur DNA having been ingested by mosquitoes, which then became fixed in amber, de-extinction projects always begin with tissue preserved by freezing or alcohol fixatives. But as synthetic biology and allied bioengineering projects look toward the future, the rhetorics with which boosters promote and describe their extraordinary visions sound increasingly like the extrapolative and conjectural thinking that animates science fiction.

A Biological Time Machine

National Geographic hosted the first public meeting of the Revive and Restore project, which was held in Washington, DC, in May 2013. Attending on George Church's invitation, I noted the event's slick production values. It was cohosted by the now-ubiquitous TED brand of conferences, which streams filmed short presentations by academic and pop-culture personalities as a facile and easily digestible blend of "edutainment" and self-help. There were more reporters and television crews crowding the periphery of the room than there were audience members populating its seats. Though the weather on the streets of the capital was muggy that spring day, the interior of the amphitheater was nearly glacial. Taking out pen and paper, I hoped my fingertips would soon thaw enough to take notes. Noodly Brian Eno and David Byrne music provided a soundtrack in between publicity-friendly fifteen-minute lectures given by zoologists, paleoecologists, molecular paleontologists, and synthetic biologists. Each scientist who took the stage that day promised that massively parallel DNA synthesis of the sort developed in Church's laboratory made resurrecting extinct species (woolly mammoths, passenger pigeons, quaggas[7]) not merely a possibility but an inevitability, one achievable in the next decade.

The idea for this conference had originated two years earlier, when George Church had dinner with Stewart Brand and his wife, Ryan Phelan. Brand founded the *Whole Earth Catalog*, which, beginning in the 1960s, joined Bay Area counterculture with the burgeoning digital intelligentsia of Silicon Valley. The catalog, which marketed goods to hippies and high-tech types, would later serve as a model for the early Internet.[8] More recently, Brand founded the Long Now Foundation, a nonprofit that cultivates technologies and products that address the deep future, such as a clock that will accurately keep time for ten thousand years. The gatherings he hosts, this one included, have a decidedly science-fictional appeal, and speakers at Long Now events include speculative fiction and futurist authors such as Neal Stephenson and Ray Kurzweil. Ryan Phelan is on the board of Church's Personal Genome Project, founded DNADirect.com (a direct-to-consumer genomic sequencing company), and directs Revive and Restore, which is a subproject of Long Now. When Church told Brand and Phelan about his intention to resurrect a woolly mammoth and a passenger pigeon, Brand and Phelan were enthusiastic, immediately offering financial support and publicity.

A *National Geographic* journalist who inaugurated the day's proceedings fantasized about a time machine that would transport him into the deep past. He imagined, after H. G. Wells, surfing across geological epochs to spy pterosaurs gliding through the Cretaceous, Eocene moeritheres wallowing in swamps, and a ponderous *Ambulocetus* clumsily waddling across dry land to wade in hot saltwater. He emphasized that while such "time travel" can currently only be approached by studying fossils, none of what he described is "purely SF; it's a proposition" that could be fulfilled by genomic synthesis.

To underline the technological feasibility of such a fantastic scheme, several speakers pointed to precursors that suggest the technology is more realistic than it might currently sound. For example, they reported on researchers who used SCNT to produce a live mouse generated from the nuclear DNA of a mouse corpse kept on ice at $-20°$ Celsius for sixteen years. They compared this feat to bringing the mouse "back from the dead."

Such language smacks not just of eschatological fantasy but also of genetic essentialism, since it assumes that nucleic DNA is capable not merely of producing a genetic chimera of the first organism but of reviving a particular biological individual as well. The 2008 *Proceedings of the National Academy of Sciences* article reporting on this research immediately leapt from frozen mice to larger game: "it has been suggested," the authors write, "that the resurrection of frozen extinct species (such as the woolly mammoth) is impracticable, as no live cells are available, and the genomic material that remains is inevitably degraded. . . . Thus, nuclear transfer techniques could be used to 'resurrect' animals."[9] They underscore the trope of resurrection by juxtaposing photographs of a very alive mouse pup beside the withered frozen carcass of its genetic progenitor.

SCNT is a technology in which an entire adult cell from one animal is inserted into an egg that has had its nucleus removed in order to make a new kind of embryo. It was pioneered in the 1990s at the Roslin Institute, now famed for the "cloning" of Dolly the Sheep, as well as her ovine kin Polly, Molly, Megan, and Morag. Church references Dolly's birth in describing his own project, proclaiming, "Genomic technologies can actually allow us to raise the dead. . . . Although Dolly's genetic parent had not been taken from the grave and magically resurrected, Dolly was nevertheless probably a nearly exact genetic duplicate of the deceased ewe from which she had been cloned, and so in that sense Dolly had indeed been 'raised from the dead.'"[10]

In her account of the making of Dolly the Sheep, Sarah Franklin recognizes SCNT as a biotechnology that effectively reverses biological time by "dedifferentiating" cells, that is, inducing adult cells to behave like stem cells. "In other words, through biotechniques, the temporality of the biological is being rescaled, or even recreated. If specialized adult cells appeared to be going back in time when they were, in fact, always already in a germinal temporality and telos to begin with (i.e., because they never left it), then it appears that one of the basic formal properties of the biological was only 'there' to the extent it was assumed it had to be."[11] Extending the temporality of mammalian cells to the temporalities of whole organisms and entire species, Church and his colleagues in synthetic and conservation biology expect to rewind and traverse millennia. They propose hybridizing not only organisms' genomes but also the chronologies species inhabit.

Lazarus on Ice

At the Revive and Restore meeting, Michael Archer, a paleontologist at the University of New South Wales, narrated the strange tale of the gastric brooding frog (*Rheobatrachus*). The gastric brooding frog's name explains exactly what it is, or what it was, or what it might soon be. The frog is so named because it once birthed live frogs from its mouth. Discovered in 1972, by the mid-1980s it was extinct, felled by a rogue fungus. Archer obsessed over reviving the gastric brooding frog, launching the Lazarus Project to resurrect it after he learned that a nearby professor at the University of Adelaide had been the last researcher to keep a colony of these frogs in his laboratory. A jar of frog tissue had been stashed in a −20° Celsius freezer in Adelaide for over forty years.

Using SCNT, Archer introduced *Rheobatrachus* nuclei into host eggs of the related barred frog (*Mixophyes*). His team watched with astonishment as embryos divided under the microscope. While there was no reason, he said, to think a "dead nucleus" would take hold in a "live species," as he put it, "I saw a miracle starting to happen." One of the eggs began to divide "again and again." These live cells contained the nucleic DNA of a previously extinct species. While none of these cells successfully reached the blastocyst stage or gastrulated (i.e., they died long before a viable frog could hatch), Archer ended his presentation buoyantly, exulting with arms raised in a flourish of old Judeo-Christian overreach: "We are watching Lazarus rise from the dead . . . step by hopping step!"

Alberto Fernández-Arias, head of the Department of Hunting, Fishing, and Wetlands in Aragon, Spain, next took the stage to offer his own, much more solemn and dispassionate presentation. He chronicled for us a marginally more successful resurrection story. As journalist Carl Zimmer would later introduce the tale, "On July 30, 2003, a team of Spanish and French scientists *reversed time*."[12] In the Ordesa and Monte Perdido National Park running along the border between France and Spain, the population of a Pyrenean ibex known as the bucardo dwindled in the late twentieth century, a victim of European hunters. On January 6, 2000, the last female bucardo, Celia, was crushed under a falling tree, rendering her species extinct. Researchers discovered her mangled body after her tracking collar set off a "mortality alarm" that was triggered when she stopped moving.

But ten months earlier, Fernández-Arias had taken a tissue sample from Celia and frozen it. In the months that followed, a team of scientists from Aragon and INRA (an agronomic research institute in France) performed a series of interspecies embryo transfers. They first mated Spanish ibexes with domestic goats to produce several hybrid ibex-goat females. They then transferred 154 bucardo nuclei (cloned from Celia) into enucleated goat eggs and implanted these chimeric embryos into forty-four hybrid surrogates. Only one bucardo was carried to term; it was born in July 2003. The newborn bucardo kid died in Fernández-Arias's arms within seven minutes; an autopsy showed a supplementary lung lobe filling her thoracic cavity. The project died shortly after the cloned kid did, as the Spanish government no longer wanted to subsidize a project to resurrect a recently extinct ibex. To date, the bucardo remains extinct, its tissue suspended in ice waiting for eventual reanimation. Nonetheless, Fernández-Arias (echoed by other scientists who spoke at the conference) is undeterred: he considers the bucardo the first success in SCNT de-extinction technology.

After another synthesizer-heavy ambient interlude, Oliver Ryder stood before us to describe his work at the helm of the San Diego Zoo's Frozen Zoo, a cryopreservation lab founded to bank biodiversity for conservation efforts. After thirty-five years, the Frozen Zoo maintains the best-characterized and most extensive stocks of frozen animal tissue in the world: it has samples from over 503 mammals, 170 birds, 70 reptiles, and 12 amphibians, all endangered, rare, or extinct. After extinction, tissue samples kept on ice at the Frozen Zoo are all the biomaterial that remains of a species. The Roslin Institute maintains a similar facility, the Frozen Aviary, which takes germ cells from birds and grows them in tissue culture before freezing them.

Researchers also use other technologies to document, archive, and catalog endangered animals, capturing the lives of animals nearly extinct. For example, popular-science articles reporting on de-extinction make ample use of photographs of taxidermied specimens: bucardos, passenger pigeons, and thylacines arranged in lifelike poses. Not least among these analog technologies is cinema, as grainy videos capture the death throes of the last of a species, frozen in time.

Archer began the Thylacine Project in 1999 while he was director of the Australian Museum. The thylacine, or "Tasmanian tiger," long endemic to Australia, New Guinea, and Tasmania, was a marsupial that roamed rainforests 25 million years ago alongside giant ducks, tree-climbing crocodiles, and carnivorous kangaroos. It disappeared from Australia centuries ago, yet ranged across Tasmania into the twentieth century. When the British arrived there, they brought herds of sheep with them and began a sustained campaign to shoot thylacines. Suspecting them of ravaging their herds, British officials placed bounties on thylacines' heads. As thylacine numbers dwindled precipitously, the bounty was lifted, and the government extended legal protection to the marsupial. Benjamin, the last known thylacine, died of exposure at the Hobart Zoo in Tasmania in 1936, just fifty-nine days after his kind had received legal protection.[13] His body was thrown into a city dump.

A black-and-white film of Benjamin, which Archer screened during his presentation, captures him pacing nervously in his fenced pen, patrolling its perimeter. The camera then cuts to him standing guard over a hunk of flesh, tearing meat from bone. Benjamin is preserved in this filmic fixative—he sniffs contemplatively at the air, edgily regards the camera, stretches and scratches at his left flank. The room was dead quiet, as everyone watched the film somberly; many viewers wore mournful expressions on their faces. Historian of science Etienne Benson reads film as one of an arsenal of archival technologies wielded by conservationists in their efforts to reconstruct endangered species.[14] In this respect, the nature film serves as "an exact image . . . to ensure against disappearance, to cannibalize life until it is safely and permanently a specular image, a ghost. The image arrested decay."[15] The thylacine, despite Archer's best efforts, remains a specter. It has yet to be resurrected.

In 1990 Archer discovered a jar in Sydney's Australian Museum in which was floating "a little girl thylacine pup." Hunters had killed her mother in 1866 and pickled the pup in alcohol (ethanol curbs DNA degradation). The jar had been squirreled away and lost in the museum archives for over a century. When Archer set about to recover DNA from the preserved specimen, much of what he found was not thylacine but human DNA, which he surmises was residue from humans handling the animal after its death. Nonetheless, he recovered thylacine DNA, amplified it using polymerase chain reaction, and established a genomic library. Archer anxiously awaits Church's work on MAGE so he may begin using the fragmented thylacine DNA to reintroduce the species into its native habitats, using the related Tasmanian devil as a surrogate species.[16] When he asked us how many audience members would want to keep a thylacine as a pet once it is resurrected, almost all the people in the room raised their hands.[17]

Cells suspended in deep freeze. Frozen mice "resurrected." Mammoths biding their time in permafrost as millennia slowly tick by. It seems biotechnical resurrection is always a frosty enterprise. Such climatic conditions enable what historian of science Hannah Landecker has described as one of the foundational tools of biotechnology: the suspension and resumption of biological time. Natural philosophers and scientists have long sought to use extreme cold to pause and reanimate life, beginning with Robert Boyle's seventeenth-century experiments freezing ox brains. In the first decades of the twentieth century, biologists learned to freeze and successfully thaw microbes and spermatozoa; in the 1950s some experimented with reanimating hamsters. In the 1960s Robert Ettinger built on earlier research freezing human tissues to pioneer cryogenics, freezing dead humans' brains in hopes that they may someday be reanimated and thereby achieve immortality.[18]

In her history of tissue culture, Landecker describes the material infrastructures that enable the proliferation of cells and tissues outside the body. Cryobiology, she writes, has long been a technique of managing biological reproduction. "All the disassembled generations, the novel simultaneities, the gaps of time between death of one generation and birth of another with a suspension of continuity between them—all of these deeply unsettling temporal disruptions depend to some degree on the rather banal presence of a working deep freeze."[19] Put otherwise, "Freezing looped the line in lineage."[20]

Salvage ethnographers and salvage biologists have for over a century been allied in their efforts to capture and preserve endangered cultures, species, languages, and ecosystems before they disappear. Bronisław Malinowski inaugurated ethnography's project in the first sentences of *Argonauts of the Western Pacific*: "Ethnology is in the sadly ludicrous, not to say tragic, position, that at the very moment when it begins to put its workshop in order, to forge its proper tools, to start ready for work on its appointed task, the material of its study melts away with hopeless rapidity. . . . When men fully trained for the work have begun to travel into savage countries and study their inhabitants—these die away under our very eyes."[21] A hundred years later, biologists bank on similar notions of the endangered "primitive" to frame their scientific enterprise and to jumpstart a sensibility in which freezing can arrest the endangered past from liquefying and "melt[ing] away."

Techniques of cryopreservation have for decades been enrolled in biological and anthropological salvage missions and in so doing often earmark those individuals and populations whose tissues are frozen as being inherently in danger or at risk. So too, de-extinction researchers mine frozen tissues "latent" in permafrost and lab deep-freezers in order to launch wildlife relegated to the past into imagined biological futures. Joanna Radin argues that Cold War biologists who froze and databased blood samples from "primitive peoples" trafficked in "ideas about the latent, or as yet untapped, genetic knowledge of humans thought to be portals to the past. These concepts were joined with future-oriented beliefs that the ability to make primitive blood physiologically latent—to freeze it—would enable it to become available for new uses even after the communities themselves had disappeared."[22] She adopts "latency" as a term that places indigenous populations in "suspended animation," simultaneously relegating them to the past while orienting biological knowledge toward an anticipatory vision of the future.[23] Jenny Reardon similarly argues that human genetic diversity research is driven by worries that genetic value, if not banked now, will be lost to future generations of scientists.[24] In this sense, it is noteworthy that Church's MAGE technology has not yet successfully synthesized any organism other than bacteria, yet an anticipatory stance toward the tools of synthetic biology promises that, as synthesis technology continues to develop, extinct animals will be resurrected in the not-too-distant future.

Further, the politics of zoological salvage missions mirror the ethical troubles of anthropological ones: like their predecessors, synthetic biologists and conservation biologists make decisions about which organisms and populations are considered the most charismatic and genetically valuable. Salvage projects assume that a population is already lost, attempting to freeze that population in the past rather than redoubling efforts to care for those people, lifeways, or organisms still very much alive, present, and at risk. For example, while paleoecologists tramp through the tundra in search of woolly mammoth tusks to use as raw material for DNA sequencing, poachers kill twenty-five thousand African elephants a year to sell their ivory tusks on the black market. The same animals slated to serve as surrogates for extinct organisms are themselves currently endangered.[25]

In the 1970s the US Fish and Wildlife Service championed the Endangered Species Act using the slogan "Extinct Is Forever." During the de-extinction conference, scientists repeatedly upended this mantra. Archer posed it as a rhetorical question: "Does extinction have to be forever?" Robert Lanza, the chief scientific officer of Advanced Cell Technology, affirmed, "Extinction is *not* forever." A wildlife ecologist declared, "If extinction isn't forever, some fundamentals change: 3.8 billion years of life on Earth comes undone." The Frozen Zoos and Aviaries, the jars of frozen frog tissue and the mice kept on ice, when paired with the temporal hybridizations enabled by SCNT and genome synthesis, turn individuals and species into living things that do not necessarily live in their own times but instead are subject to temporal discontinuities, disruptions, nonlinear lineages, chronological graftings, and seemingly mystical resuscitations.

Coldness synthesizes and reassembles, queers and hybridizes biological time. The resulting organisms are, to borrow the language of Dipesh
Chakrabarty, a sort of "timeknot" that "refer[s] us to the plurality that inheres in the 'now,' the lack of totality, the constant fragmentariness, that
constitutes one's present."[26] Carolyn Dinshaw reminds us that such "asyn-
chronous" or heterogeneous presents, in which "different time frames or
temporal systems collid[e] in a single moment of *now,*" are "queer by virtue
of their particular engagements with time."[27] Just as DNA synthesis allows
synthetic biologists to hybridize a rock pigeon with a passenger pigeon genome, or a woolly mammoth with an African elephant genome, freezing
does not simply pause, reverse, or restart biological time but grafts the past
(species whose time has run out) and the past imperfect (rare or endangered species) onto the present (species filling in as surrogates for rare animals) in order to capacitate future life-forms.

Speeding Up Biological Time

Church describes synthetic biologists' De-extinction Project using a phrase
that recalls Ernst Haeckel's biogenetic law that maps embryological onto
evolutionary time. For Haeckel, "ontogeny recapitulates phylogeny,"
meaning that the morphogenetic development of an individual organism
in utero or in ovo follows, in miniature and sped up, the successive stages
of evolution that its ancestors underwent over the course of millennia.
Church forwards his own biogenetic law: "engineering recapitulates evolution."[28] By this he means that synthetic biology has begun by engineering
simple life-forms (viruses and bacteria) but will quickly proceed to manufacturing higher life-forms: mammals and even humans.

Church and fellow synthetic biologist Peter Carr together developed
and patented the MAGE technique, a synthesis method that allows synthetic biologists to make up to fifty allelic replacements at a time, producing billions of genetic variants ranging from a few nucleotides to entire genes in a few hours' time.[29] They tout MAGE as one of a series of
next-generation synthesis technologies that will vastly accelerate DNA
synthesis and genome-level engineering at a relatively low cost (on the order of thousands of dollars per genome). While to date its inventors have
used MAGE only to modify bacterial plasmids, they propose that one
application of this technology would be "speed[ing] up" or "accelerating"
directed evolution,[30] by which they mean multiplying genetic variations,
effectively making billions of genetic modifications and then selecting for

desirable characteristics among them (e.g., greater output of a desired protein or chemical). Such forward engineering is a tool that allows genomicists and bioengineers to sift through randomly evolved organisms for a sought-after trait rather than using rational design principles to engineer a new organism exhibiting that characteristic.

Another application Church proposes using MAGE for is rapidly introducing multiple genes from extinct species into their closest living relatives. Foremost in Church's mind is resurrecting woolly mammoths and Neanderthals, whose genomes would be overlaid onto (and whose embryos would be gestated in the wombs of) elephants and humans, respectively. In January 2013 the media blasted Church after an interview with him in *Der Spiegel* publicized a quotation from his recent book, *Regenesis*, in which he bragged that DNA synthesis had progressed to a point at which it was technically feasible to clone a Neanderthal. Church writes that he merely needs an "adventurous human female" to serve as a surrogate.[31] After a flurry of bad press, Church dialed back the claim, stating that the quotation was taken out of context and poorly translated from English into German. Cloning a woolly mammoth sounds like a stunt, and perhaps it is, but Church's ultimate aim is speeding up the reconstruction of whole genomes to rebuild new species, as he puts it, "from information alone."

MAGE is one example of how the rhetoric of acceleration operates in biotechnology to imagine a future in which biological manipulation can be accomplished faster than it is now. Synthetic biologists often cite the "Carlson Curve," which graphs an exponential increase every eighteen months in the speed of DNA sequencing and synthesis and a concomitant exponential decrease in the cost of these technologies. The Carlson Curve is the biotechnical equivalent of Moore's Law for digital technology. Like the Carlson Curve, faith that MAGE will accelerate genomic engineering is "a mode of historical consciousness that privileges the inevitability of technological progress over the inevitability of human power."[32] Indeed, anthropologists and historians of science have argued that speed and acceleration are tropes that generate revenue in the biotech industry.[33]

Examining de-extinction efforts reveals how synthetic biologists join ecological futures with endlessly recombinable biological pasts, rhetorically transmuting species lost into life-forms to come. De-extinction researchers are certainly invested in expectations of how biological species revived by accelerating biotechnologies might generate future value. However, they are also deeply invested in the *past* rather than the future and

derive biotechnical value from stances that are nostalgic rather than speculative. Anthropologists, sociologists, and historians of the life sciences
have, in the last decade, turned their attention to biocapital—in particular,
how future-oriented thinking generates hype and profit in the biotechnical
present.[34] Speculative or promissory thinking in the life sciences—about
new biotechnologies, pharmaceuticals, companies, and cures—builds upon
the generativity, vitality, and potency of biological substance.[35] Synthetic
biologists not only focus on profits generated from the hoped-for vitality
of living substance but also bank—in both senses of the term—on biological species in the *past tense*, those organisms already extinguished.[36] To riff
on both the thermal and the fiscal aspects of this enterprise, this may be
thought of as a form of biological "liquidation," in which a frozen biological past is, through its endangerment (Malinowski's worries over heritages
"melt[ing] away"), converted from a genetic asset into an object with cash
value.

"Proxies, Fakes, Theater Props"

At the 2013 meeting, Robert Lanza described to us the first instance of
interspecies nuclear transfer that resulted in a live birth. An endangered
Southeast Asian wild ox (*Bos gaurus*) was born in Sioux Center, Iowa, in
January 2001. Researchers mailed frozen gaur embryos, made by inserting gaur nuclei into cow eggs, to a farm in Iowa, where several unassum-
ing cows were slated to be impregnated. After a few false starts—FedEx
forgot to pick up the package and the embryos had thawed by the following morning—researchers artificially inseminated four cows with gaur
embryos. Of the four resulting pregnancies, one cow, Bessie, birthed a live
gaur. Lanza admitted that the gaur, named Noah, lived for only two days.
While the official cause of death was dysentery, Lanza offered by way of
explanation that gaurs are "usually born in the bamboo jungles of southeast Asia [while this one was] born on an Iowa farm."

    What defines a species as a biological object, distinct and insulated
from other species, rather than a grade of organisms existing along a
cline, promiscuously swapping DNA with other creatures? Interspecies
SCNT, like new reproductive technologies in humans, upsets regnant notions of proper kinship, but SCNT does so by drawing together organisms
classed as *too* different. Extending cultural kinship to biological relatedness, anthropologists studying assisted reproductive technologies (ARTs)

have pointed out that cloning is a taboo form of reproduction because it involves "too much sameness" or "too much homosex."[37] Extending this claim, interspecies surrogacy is an illicit example of "not enough kinship"—a sort of kinky, illegitimate, and artificial procreation across species boundaries.

A second technology that conservation biologists are pursuing to resurrect extinct species highlights the problem of species purity and in particular the ways in which evolution and natural selection lead to speciation. Selective backbreeding is a sort of artificial selection that plays out in reverse: researchers aim to transform a modern-day species into an extinct ancestral species. Looking at drawings and illustrations of primitive cattle breeds, they attempt to bring all the phenotypic characteristics of an extinct breed into successive generations of living cattle. Henri Kerkdijk-Otten, curator of the Prehistoric Megafauna Fossil Collection at the World Museum of Man, is on a singular mission to revive a wild prehistoric bovine called the auroch, the last of which died in Poland in the seventeenth century.[38] Working with old cattle breeds, researchers today aim to slowly breed cattle so that, over the course of multiple generations, the cattle sport all the characteristics of what Kerkdijk-Otten terms an "original auroch." The method involves artificially inseminating bison to create crossbred Holstein cattle, with the goal of eventually breeding a herd of heretofore-extinct aurochs. He hopes to introduce these new aurochs into areas in which aurochs once roamed. But what does he mean by "original"? Or for that matter, what does he mean by "auroch"?

At the end of Kerkdijk-Otten's presentation, Stewart Brand again took the stage to ask Kerkdijk-Otten a pointed question: "How do you know from backbreeding when you're finished, that you have an auroch?" To rephrase Brand's question, what makes an auroch an auroch rather than just a Holstein-bison hybrid that, if you squint just so, sports some of the phenotypic characteristics of a prehistoric bovine? Kerkdijk-Otten's answer was revealing because he assumed that species is not a biological type grounded in successive and continuous generations sharing common lineage. "Instead of just taking an end product, we are gradually transforming an animal into an *original.*"

But an original what? Brand pressed the issue, sardonically asking the audience as Kerkdijk-Otten left the stage, "Is a new passenger pigeon a *real* passenger pigeon? Or are they proxies, fakes, theater props?" So too, the *National Geographic* article reporting on de-extinction featured an artist's rendering of a passenger pigeon taking flight, captioned, "If it looks and

flocks like a passenger pigeon, is it a passenger pigeon?"[39] Elsewhere, the
same author asks, "Even if Church and his colleagues manage to retrofit
every passenger pigeon–specific trait into a rock pigeon, would the resulting creature truly be a passenger pigeon or just an engineered curiosity?"[40]

Now is certainly not the first time ideas about ancestry, heritage, and
bloodline were projected onto the animal world. In the eighteenth century, British stockbreeders valued Chillingham cattle as analogous to ancient "native" British bovines and thus "identif[ied] . . . these allegedly
primordial cattle with the modern human inhabitants of their island; they
were, as one nostalgic writer put it, 'ancient Britons.'"[41] Further, species
has been a problematic category for life scientists for a long time, as, for
example, when readers of Darwin first realized that, if Darwin was right,
then species "were no longer facts, or evidence of categories of God's mind,
but judgments, perhaps even psychological projections 'made' by the tax-
onomist."[42] The definition of "species" most often used as shorthand in
biology today was first articulated in 1942 by evolutionary biologist Ernst
Mayr, who defined species as a reproductively isolated and compatible
natural population.[43]

Yet there are now multiple competing and overlapping biological definitions of "species" that vary between evolutionary and taxonomic criteria, sexually and asexually reproducing species, even definitions specific
to biological subdisciplines (e.g., ornithology vs. botany). These disputes
already clue us into the fact that "species" is definitionally unstable. If
Darwinian natural selection privileges genealogy as the marker of species, and twentieth-century molecular biology has translated genealogy
into a genetic idiom, de-extinction could compromise shared evolutionary history as a ground for biological taxonomy, reviving physiology and
behavior as principles on which to judge verisimilitude. An animal that
"looks and flocks like a passenger pigeon" *is* a passenger pigeon, even if it
harbors rock pigeon DNA and is taught to flock by rock pigeons.

Carrie Friese argues that recent SCNT techniques mark "a shift in
selective breeding from a focus on the phenotype to the genotype. The
focus is no longer on producing certain types of animals that exhibit certain phenotypic traits, but rather on producing certain kinds of genomic
configurations."[44] On the contrary, it is apparent that while genetic value
remains significant to synthesis-enabled conservation efforts, the decision
as to whether a genetic chimera counts as "real," for de-extinction scientists, rests on purity of phenotype (morphology, behavior), which trumps
genetic equivalence or continuous lineage. Indeed, since the advent of

genetic sequencing, "species" has been understood as a biogenetic category rather than a morphological one. Yet both backbreeding and Church's proposed use of MAGE synthesis to compose hybrid genomes uncouple biogenetic from morphological definitions of species. When biological origins are undone and evolutionary time is run backward, species identity is rendered problematic. There is no *there* there. What counts as "real" or "original" no longer makes any genetic, genealogical, ontological, or historical sense.

In her feminist science-fictional exploration of interspecies surrogacy and new reproductive technologies, Charis Thompson reflects upon the joined aims of human reproduction and animal conservation. In "Confessions of a Bioterrorist," Thompson imagines Mary, a lab technician at a West Coast frozen zoo, choosing to perform an act of "bioterrorism" by stealing frozen bonobo embryos and impregnating herself with them in a transgressively queer technoscientific virgin birth (it is noteworthy that Mary's due date is Christmas Day). Thompson's story makes clear that biotechnical conservation efforts trade on "a genetic conception of 'species.' And it [is] a prospector's dream come true, banking genetic diversity for posterity/prosperity."[45] Chimeras and hybrids, she suggests, are not coded as "natural," and therefore not recognized as worth saving for those who believe in a nativist, genetically grounded, and static conception of "species." Friese similarly argues of endangered animals reproduced via SCNT that "chimeras not only exist betwixt and between species. These biological organisms call into question the very notion of species itself."[46]

The ways in which old technologies (breeding, freezing) and new technologies (genomic sequencing and synthesis, interspecies surrogacy) trouble breed boundaries and species purity and undo the temporal logics by which such categories are defined were enunciated by conservation biologist Kent Redford. At the front of the *National Geographic* auditorium, he baldly stated that species conservation reflects an "overarching human desire to make categories while making things that override those categories."

As examples of such thinking, he described bison conservation efforts in the last few decades. Ranchers trying to breed bison that need less water and can survive harsh winters had for decades bred bison-cow hybrids, animals that looked like bison yet harbored cattle genes. When the Bison Society began restoring bison to American nature preserves in the 1970s, however, they resisted releasing bison that were "tainted by cattle genes." What does it matter, though, Redford asked? No species is ever "pure."

Wolves bear dog genes; humans are well stocked with Neanderthal and Denisovan genes, not to mention plenty of viral DNA. As he put it, "purity isn't found in nature, in species. When we ask why things aren't pure, we should be looking in the mirror at ourselves." His point was that at the genetic level, species always commingle.

Synthetic biologists, in seeking to resurrect extinct species, must first contend with "species" as something that is never isolable, pristine, or static.[47] Instead, in the absence of genomic homogeneity, synthetic biologists are reviving old ideas about phenotypic taxonomy as a proxy for evolutionary definitions that rely on biogenetic sameness and temporal continuity. Organisms that look and behave like an auroch, a wolf, a bison, a passenger pigeon, or a human simply get coded as such.

Pleistocene Park

The issue of what constitutes a "species" is a live one for George Church in his efforts to resurrect the woolly mammoth. Rather than working with the entirety of a woolly mammoth genome, he proposes hybrid genome reconstruction. The idea is as follows: remains of woolly mammoths, that most charismatic of Pleistocene megafauna, are frozen in permafrost across the Siberian tundra. Beth Shapiro, who collects mammoth carcasses in the Arctic, flashed photos across the screen of mammoth teeth, bones, and hair, even massive heads bearing tusks and still encasing ancient proboscid brains. Frozen tissue preserves DNA, even as bacteria slowly metabolize and fragment it. Working backward from frozen mammoth tissue collected in the Arctic and gold-mining sites in Alaska, biologists are currently compiling sequences of mammoth DNA. The first sequence of a woolly mammoth genome was published in *Nature* in 2008, the concerted work of scientists from Pennsylvania State University, the Severtsov Institute of Ecology and Evolution, the Zoological Institute of the Russian Academy of Sciences, the University of California, Santa Cruz, the Broad Institute, and Roche Diagnostics. Hair samples from two woolly mammoths, frozen in permafrost for twenty thousand years, yielded 4.17 billion bases of nuclear genome sequence, most of it positively identified as mammoth in origin.[48]

Some biologists are now combing Siberia for mammoth tissue that still contains a living frozen cell or nucleus that could be ported into an enucleated elephant egg. Foremost among these teams are researchers at the Sooam Biotech Research Foundation in Seoul, South Korea, who are working alongside biologists in Yakutsk, Sakha Republic. The researchers who presented in DC instead proposed continuing to use high-throughput DNA sequencing to compile fragments of mammoth DNA from extant tissue, which is much easier to procure and more likely to be viable than whole cells. They would then stitch those genetic fragments together until they could compose the full complement of the mammoth genome. Comparing that genome to the mammoth's closest living relative, the Asian elephant, synthetic biologists would synthesize those portions of mammoth DNA not shared by both species, then use Church's MAGE or some other synthesis technology to port those fragments into an elephant genome. Essentially, Church wants to insert genes that would code for mammoth hemoglobin, change the hair color of the Asian elephant, increase its hair density, amp up its subcutaneous fat and sebaceous glands, and increase its tusk length. The majority of the genome would be elephantine with a few key genes coding for mammoth physiognomy mixed in.

The next step, SCNT, would allow Church and colleagues to insert the hybrid mammoth-elephant genome into an elephant ovum using microinjection. They would then induce it to divide via chemical means or an electric stimulus. The cell would need to divide and reach the blastocyst stage before being introduced into the uterus of a surrogate elephant that would bring it to term. A biologist in the audience pointed out that woolly mammoths were much larger than Asian or African elephants, and she was concerned that the surrogate elephant would be uncomfortable, if not endangered, during the twenty-two-month gestation period during which the surrogate would bring a mammoth to term.

While much of the mammoth genome has already been sequenced, the rest of the project remains highly speculative and prone to starry-eyed, boyish *Jurassic Park* scenarios. Kerkdijk-Otten fantasized that his ecotourist destination, a wildlife park he names "the Danube Delta," would offer visitors an experience similar to the safaris of the African Serengeti. He imagines herds of wild buffalo, horses, and red deer roaming free across the alluvial plains and wetlands of Romania and Ukraine. Another evolutionary biologist ended his presentation by nominating large swaths of Siberia and the Yukon as appropriate habitats for the resurrected mammoths, declaring, "the boy in me wants to see these animals walk across the permafrost."

As Church told me when we spoke in his office, he hopes that resurrecting the woolly mammoth will alleviate some of the effects of anthropogenic climate change. This idea was first promoted by Sergey Zimov, a geophysicist from Vladivostok who currently directs the Northeast Science Station in Cherskii, Sakha (Yakutia) Republic. In 1989 Zimov proposed restoring a Pleistocene ecosystem that, for over a million years, had spanned the steppes and tundras of northern Europe, northern China, Siberia, and, across the Bering Strait, Canada. Zimov writes that in Siberia "[m]ammoths, woolly rhinoceroses, bison, horses, reindeer, musk oxen, elk, moose, saiga, and yaks grazed on grasslands under the predatory gaze of cave lions and wolves."[49] This ecosystem disappeared ten thousand years ago. In a subsection titled "The Future of the Past" in an article in *Science*, Zimov reports on efforts to restore this ancient ecosystem on 160 square kilometers of land in the Sakha Republic. Many mammals that walked, tramped, hopped, or galloped alongside the woolly mammoth remain in Siberia, and in DC Zimov announced his plans to reintroduce bison and Siberian tigers to the land. Church then hopes to populate the land with herds of resurrected mammoths. In a Crichtonesque flourish, the resulting ecosystem is to be named "Pleistocene Park." Large stores of carbon are sequestered in the Arctic permafrost, and as the tundra thaws, this carbon is released into the atmosphere. Church imagines herds of woolly mammoths roaming the Arctic tundra, their massive feet trampling winter snow to maintain the permafrost and cold temperatures.

The "denial of coevalness," Johannes Fabian suggests, generates knowledge of human Others by secularizing and *spatializing time*, drawing upon evolutionary theories to think about contemporary ethnographic subjects as existing in a time other than the present.[50] The proposed park is being cultivated on land currently populated by 956,000 people, about half of whom are ethnic Yakuts. Yet, by populating this space with "primitive" species hailing from another millennium and dispensing with evolutionary history, Zimov and Church temporalize space, coding present and future ecologies as already antecedent.

Church writes, "It would be the closest thing to time travel: a return to the flora and fauna of the Pleistocene epoch, a sort of latter-day Siberian Eden."[51] Describing Pleistocene Park as a latter-day Eden effectively erases the technological interventions, ecological crises, and anthropogenic extinctions that would give rise to this attraction. A biblical Eden assumes a state of untouched nature prior to human intervention. Logics of biological conservation and restoration envision engineered environments that reconstruct primeval "wildernesses" as potential utopias. But as environmental historian William Cronon points out, "there is nothing

natural about the concept of wilderness." Euro-Americans' Romantic at-
tachments to wilderness depend upon construing constructed places as
natural and beyond civilization's reach. Furthermore, "wilderness represents a flight from history. Seen as the original garden, it is a place outside
of time, from which human beings had to be ejected before the fallen
world of history could properly begin."[52]

In stocking modern-day Sakha Republic with the revived zoological
representatives of a distant epoch, De-extinction Project participants seek
to expiate anthropogenic ecological destruction and species extinction by
leveraging the central gambit of synthetic biology. Namely, synthetic biologists construct synthetic life-forms and then treat them as if they are
not mere stand-ins for nature but exemplars of what biology has *always*
been. Synthetic nature here masquerades as a wild primeval landscape;
life-forms not yet made become props for creatures already long gone.

"Tick, Tick, Tick"

At the heart of each of these stories is a fantasy of resurrecting the past.
Biologists imagine raising extinct animals from the dead, looking skyward
to see the sun eclipsed by millions of swooping passenger pigeons, putting
a collar and leash on the descendant of an animal who, the last of its kind,
had paced frantically in a Tasmanian zoo eighty years ago, hearing the call
of a woodpecker whose song is now embalmed in a scratchy recording.

In teaming up with conservation biologists, synthetic biologists seek
to span million-year gaps in biological history. They aspire to effectively
rewind phylogeny, to extract species from the times in which they grew
and flourished and waned and died and to breathe new synthetic life into
them. In so doing, species become differently attached to the chronologies by which they are defined. Synthetic biologists are rewriting species
as a category that, like Shirley Temple Three, is "yanked out of its own
time."

Synthetic biologists' resurrection and de-extinction projects are fantasies
of ultimate biological control: their interventions seek to produce wholly
synthetic creatures that will stand in, counterintuitively, as semblances of
untouched "nature," a latter-day Garden of Eden seemingly unsullied by
human hands, albeit generated by the most recent bioengineering technologies. Such thinking is buoyed by synthetic biologists' ambivalent stance
toward their own status as "creators" of synthetic life, with all the hubris

and hype such biblical thinking generates. If biological manipulation and modification serve as a benchmark of biological mastery and operate as tools by which synthetic biologists seek to study life, de-extinction efforts push the boundaries of what counts as "life" and "species" by remaking them as kinds no longer bound by the temporal contingencies of death and extinction. Instead, life is rendered "interruptible,"[53] temporally and materially hybridized, and always available for reanimation.

In fantasizing about forging synthetic wildernesses that mimic ancient ecosystems, synthetic biologists also articulate possible utopias that promise to turn nature preserves into tourist attractions. If synthetic biologists seek to better understand life through its reconstruction, de-extinction promises to reconstruct a kind of life that is set loose from its chronological moorings, to reinvent species and kind as marked by phenotypic verisimilitude rather than genetic purity to type, and to generate engineered spaces coded as ahistorically "natural."

T7.1 and de-extinction could be read, perhaps, not as antithetical stances toward biological design (one deleting evolutionary pasts, the other resuscitating them) but instead as shared efforts to reform, refashion, revive, or reanimate a biological past to conform with expectations about a biological present and apprehensions about biological futures. Marx reminds us that revolution is always haunted by a lurking historicism: "And just when they seem engaged in revolutionizing themselves and things, in creating something that has never existed . . . they anxiously conjure up the spirits of the past to their service and borrow from them names."[54]

The object of synthetic biology here is making new sorts of life that dispense with any meaningful ground for what is "biological" or "natural" in favor of organisms and spaces that merely appear *as if* natural. Synthetic biologists use coldness to rhetorically hybridize evolutionary time, and thereby undo "species" as a coherent genealogical and historical category. Researchers frame the biotechnologies capacitating de-extinction efforts as forms of biological time travel that loop past, present, and future. They thereby manage biological time as something that can be stopped and started, which can stutter, pause, and sometimes reverse.

Places like Pleistocene Park would be made possible by professional and popular nostalgias for those species considered "pure" and "primordial" and therefore valued as worth reviving. Paired to this conception of an untouched biological past is a resonant anxiety about the limits of a biological future, although efforts to maintain species diversity and care for endangered populations—those animals not yet dead—are eclipsed in these narratives by the transgressive possibilities and technological wonderment of bringing back creatures already dead. Such an anxious forward-looking nostalgia is, perhaps, the signature "off-modern" sensibility of our time.[55] As one biologist at the conference put it, "Tick, tick, tick. Time's running out."

Conclusion

Allow me to end by circling back to when we began, or sometime close to it. The summer of 2005, I sit in Endy's office as the afternoon sun streams in through windows overlooking MIT's Stata Center, a building whose design has been criticized by architectural theorists for reversing the sort of computational algorithms that otherwise generate simulations of living form.[1] It is a fitting prospect for someone driven to abstract, formalize, and fragment the stuff of life. Endy taps a bit on his keyboard, then swivels his desktop screen toward me so I can contemplate an image of a well-known painting, René Magritte's *Clairvoyance (Self-Portrait)*, from 1936 (fig. 7.1).

The painting depicts, in earthy browns and eggshell blues, an artist seated at his easel, applying brush to canvas while gazing intently over his left shoulder at a single white egg, which is poised on a draped table. He paints not the egg, but a gray bird, wings extended in flight, claws delicately tucked beneath its downy body. Endy looks over his shoulder to me, gesturing toward the screen on which appears the image of the artist, who is himself turning away from his canvas. From my perch on the sofa, I am reminded of the Droste effect, the recursive nesting of one image inside another, named for the box of Dutch cocoa powder. Endy often meditates upon Magritte's surrealist painting because it helps him think about the aims of synthetic biology.

He would later write brief comments on his laboratory's editable online lab notebook, explaining:

> At least three models could explain the scene. First, the man is clairvoyant—he is able to perceive the potential of the egg and paint the appropriate animal (i.e., bird in place of platypus). Second, the man has prior knowledge that eggs

FIGURE 7.1. René Magritte (Belgium, 1898–1967), *La clairvoyance (Self-Portrait)*, 1936, oil on canvas, 21 1/4″ × 25 9/16″ (54 × 65 cm). Mr. and Mrs. Wilbur Ross. © 2016 C. Herscovici / Artists Rights Society (ARS), New York.

of a certain type turn into particular birds—he describes what he expects will occur based on past experience. Third, the man has the ability, hidden from the viewer, to determine the relevant physical state of the egg. Furthermore, the man has access to a "standard model" for cellular chemistry and physics. Taken together, he is able to observe any particular egg and predict the relevant properties of the resulting animal.[2]

In his interpretation, a deductive explanation is sandwiched between a vitalist and a mechanist one. Magritte's painting, and Endy's analysis of it, condense synthetic biology's project as one that tethers modes of reasoning to ways of animating new life-forms. His interpretation of *Clairvoyance* maps out the way of theorizing life that distinguishes synthetic biologists from other life scientists. The first model, which Endy (alongside Magritte) labels "clairvoyance," is a *vitalist* account of a bird already coiled within an egg's "potential." Its outcome is legible to the artist, who can therefore be predictively descriptive.

Endy's second construal, that the painting is based on prior knowledge of eggs and their relation to birds, is a *deductive* mode of reasoning about life, the sort of thinking that drives laboratory experimentalism. Prior observation and trained judgment, on this view, are enough to identify the relevant bird that will hatch from any given egg and, further, to surmise its ontogenesis. It is this sort of deductive, analytic method that synthetic biologists hope to supersede by replacing experimentalism with manufacture.

And it is in this respect that Endy ascribes his final interpretation— that there is nothing to be known about the egg apart from physicochemical formalisms—to the project of synthetic biology. This was the central fantasy and failure that inspired MIT synthetic biologists to design bacteriophage T7.1: the "standard model" was *not enough* for Endy to "predict the relevant properties of the resulting" phage.

Indulge me, however, in my own second-order observation, a different reading of a reading of a painting about painting. Rather than the painting encapsulating deductive or mechanistic modes of reasoning about birds and their problematic relation to eggs (with all the metaphysical feedback loops that particular axiom triggers), perhaps it instead suggests a *constructive* mode of reasoning. That is, researchers assemble the stuff of life in order to generate newly materialized biological theories. This approach might also be a *projective* or anticipatory sort of reasoning: only the egg exists, but the bird promises to materialize in a different medium by the hand of the artist (here, an uncanny correspondence to optimistic efforts to resurrect passenger pigeons is unmistakable). Or in a stronger sense, perhaps the sort of bird that will hatch from this egg is now beside the point, because the artist has now painted his own bird, which need not dovetail with the egg under observation.

The approach to making and knowing life that was inaugurated by synthetic biology inhabits a different sort of epistemic space than other methods typical of the life sciences. It is decidedly *neither* deductive nor inductive. It is indebted to but not continuous with collection, classification, and experimentalism. Nor does it align with the sort of practical design and manufacture typical of either engineering or biotechnology. Rather, it is *interested making*, making with a question attached. The objects of that making incarnate and exemplify answers, even as they point toward as-yet-unasked questions. These objects of synthetic biology are persuasive. They do not simply exemplify what life is or what it may next become. Rather, they function *synecdochically* (note: decidedly not literally or technically) as vital manifestations of theories that convince and compel synthetic biologists to redefine "life itself" as a much broader, yet nonetheless malleable, entity. Life, as I have been learning for a decade, is something that gets *made*.

Synthetic has journeyed into the epistemic mise en abyme in which theories about life nest, Matryoshka-style, within little bits of life. An ethnography of science is by necessity partial, incomplete, and fragmentary. As you hold this book in your hands, it is already outdated. The stories begun in *Synthetic* will continue to unfold and outpace my prose. Even as I draft these final pages, new developments in synthetic biology beg for careful attention and renewed analysis. In recent months, the CRISPR/ Cas9 system has made headlines: the new technology allows genomicists and bioengineers to modify an organism's germ line at an unprecedented scale. Some scientists now demand a moratorium on the technology; other labs have begun efforts to edit germ lines in human embryos. Whatever the regulatory fate of CRISPR, it heralds a fundamental change in the techniques with which genetic material will from now on be engineered.

What, then, will synthetic biology become in years to come? Will synthetic biology go the way of standardized genetic fragments, a useful technology that is now prosaic and dispersed in laboratories around the world? Or will it suffer a fate similar to that of Celia the bucardo and her cloned kin, doubly and catastrophically extinct? I expect that, whether it is a success or a failure (by whatever metric), synthetic biology will soon cease to be remarkable. This is all the more reason to attempt to make sense of it now, in its unfinished form, before synthetic biology sinks into either generalized common sense or nonsense. Either way, it will lose the veneer of strangeness—and possibility—that now dusts it.

As an ethnographer, I have worked hard not to celebrate or condemn what synthetic biologists do, let alone allow myself to guess at what they may do next or propose how they might go about doing it (whatever "it" may be) better, more safely, or more ethically. People often ask me what I think synthetic biologists might do next. I understand the curiosity behind the question, but whether synthetic biologists' desires are realized or not, their wildly ambitious venture is a parable for our present day. Lorraine Daston and Peter Galison speculate that science has recently transitioned from an age of scientific *representation* to one of *presentation*, in which the acts of "making and seeing are indistinguishable."[3] If technoscience ubiquitously structures how we grasp and intervene into the natural world, then in an age of scientific "presentation," knowing the world is reflexively and iteratively embedded in and transposed into acts of making it. This is increasingly apparent in fields ranging from nanotechnology to artificial intelligence research to econometric predictive modeling.

Synthetic is descriptive and analytic: it interrogates the first fifteen years of synthetic biology (which map onto what scholars of the life sciences term "the postgenomic moment") and elucidates how synthetic biologists have installed social forms of life—theories about design and creation, kinship and relatedness, property and exchange, labor and economy, expertise and political action, novelty and species—into the things they make. More so, it demonstrates that theories of the biological underwrite theories of the social. Though suturing *practica* to *theoria* may seem to threaten relativism, materialized theories mutate and divide and infect and colonize and, in short, behave surprisingly otherwise. Theories are lively, and life generously lends itself to theorizing.

I find synthetic biology endlessly fascinating because it is diagnostic of three things: one about life, one about science, and one about the history of life science. First, "life" is *already* a troubled and troubling epistemic category. Synthetic biologists treat life as coherent and stable despite the fact that the very thing on the table is its material and theoretical reconstruction. Second, I believe—and hope that other ethnographers will test this claim—that the story *Synthetic* tells is not limited to biology but is instead endemic to the contemporary sciences.

Third, the brave new organisms of synthetic biology, in all their gloriously messy materiality, have taught me something about the history of life's limits. They endure in bacterial plaques, lustily thrive in yeasty brews, flourish in warm rooms, bide their time in freezers, and hitch rides in test tubes and on strips of paper. Whether tamed or standardized, fragmented or refashioned, they necessarily remain corporeal and organic *stuff*. This very fact is a refreshing antidote to prior efforts to synthesize, simulate, or "make" life.

Historians of biology have appraised the menagerie of unlikely objects made by designers who, in seeking to generate theories of life, have time and again worked to simulate living processes in nonliving things. A buzzing and whirring line of lifelike precursors approximate life by aping living processes in nonliving matter: automata that play chess, serve tea, and sigh; robots that crawl, walk, and mug for the camera; Artificial Life programs in which "fish" sink or swim. They bear the features that, in their time, are most closely associated with life: metabolism and responsiveness, growth and locomotion, evolution and reproduction. Such objects do not point to an ahistorical definition of "life" but rather richly demonstrate what qualities designers consider most "lifelike" in a given historical moment.

Perhaps this history is rooted in a lurking Platonism. Substance is immaterial to an entity behaving "as if" alive; form is privileged over matter. Diverse efforts from antiquity to now are united in assuming that lifelike *processes* can transpire in *abiotic media.*[4] As Jessica Riskin summarizes, artificial life "encompasses creatures made of wood, metal, stone, glass, papier mâché, electronic circuitry, silicon, and information."[5] Life becomes a performance.[6] Reversing this trend, synthetic biology focuses on living substance, even as it often forces biology to exhibit qualities it might not otherwise have. Such creatures upend the historiographic assumption that theorizing life always relies on lifelike animacy and privileges biological matter above biological form.

Biological knowing and biological making, paired under the twin signs of the factual and the artifactual, have been mutually constitutive for some time, but that fact is now manifestly evident in the way life scientists think, talk, and work. More than pointing to an absence of any coherent or fixed referent for "life," this trend demonstrates that not only is "life itself" something richly constructed but so is biological substance, and so is biological practice. As "life itself" deliquesces and recrystallizes, reforms and deforms, in the hands of these contemporary life scientists, the relations between making and knowing are also radically reconfigured, with serious consequences for the stories we tell about nature and artifice, analysis and synthesis.

To listen to such stories and weave new ones is vital.

A Note on Method

What becomes of life when biology is understood not through experiment but through manufacture, when novel biosystems are fabricated as crucibles in which to try the accumulated store of knowledge experimental biology has produced in the last century? Throughout *Synthetic*, I have anchored questions such as these in ethnographic specificity. Rather than offering a snapshot of the field, either as it was in its origins or what it has become, my own ethnographic experience has tacked alongside the field's evolution.

I begin in 2005 at MIT, where synthetic biology was then incubating. My anthropological arrival scene—the sort where the anthropologist, surrounded by gear on a tropical beach, watches as a dinghy disappears over the horizon[1]—is not much of a scene at all. Beginning that year, I walked several times a week from MIT's East Campus, whose buildings house social sciences and humanities departments, to West Campus, the province of natural scientists and engineers. A sign along Amherst Street, on which one arrow points towards "Calculus," the other towards "Real People," marked entry to my field. While I surmised that I must count among the "real people," I made my brief westward journey each day to learn more about other real people who happened to fall on the other side of MIT's tongue-in-cheek two-culture divide.[2] Yet the two cultures had already coalesced (not that they ever had been divided in the first place).

I first met Drew Endy in his office the summer he hosted the first synthetic biology conference, a modest affair in which a few bioengineers met to lay out the groundwork for the new field of study they hoped to build and institutionalize. Endy had already received some press attention for running a winter research program in 2004 in which undergraduates genetically engineered bacteria to "blink" on and off using genes that temporally regulated expression of fluorescent proteins. These students had named themselves the "PolkaDorks." Standing in the doorway of his MIT office, I described to him in broad strokes what anthropological fieldwork is. I had grossly underestimated him. Pulling a paperback copy of Paul Rabinow's *Making PCR* from his bookshelf, he told me that if this was the sort of book I wanted to write, then I had full access to his laboratory. I began attending laboratory meetings the same afternoon.

Endy was one of my primary interlocutors during my early years in the field, and as such his shadow looms large over the first chapters of this book (and he returns at regular intervals in subsequent chapters). His recurrent presence in these pages is intentional—while this is decidedly not a book about Endy, his name was nearly synonymous with synthetic biology at this time, and in many ways his career trajectory and persona plotted alongside synthetic biology, both he and the field being earnest, even zealous, recent transplants from engineering into biology.

I conducted most of my fieldwork over the next four years in laboratories in MIT's Department of Biological Engineering and the Computer Science and Artificial Intelligence Laboratory (CSAIL). These two laboratories together made up the MIT Synthetic Biology Working Group, which from 2005 to 2009 was at the forefront of the burgeoning synthetic biology movement. The first thing I learned as I began talking to MIT synthetic biologists is that they were united in their conviction that living substance can be rendered partible and standardized and, further, that "true" bioengineering can and must follow the same principles that guide the engineering of inorganic things. Living substance, they claimed, was no different than the raw materials from which an engineer designs a bridge or an electronic circuit. Following engineering terminology, they named the three principles to which they adhered "abstraction," "decoupling," and "standardization."

Their faith in these three tenets was so pivotal to synthetic biologists that when I first asked Endy what drove synthetic biology, he gave me a lesson, not in biology, but in mechanical engineering. Endy talks animatedly, almost frantically, as if he perpetually has five minutes in which to deliver a ten-minute pitch. I would later get acclimated to his locutionary impatience, but on my first day I struggled to keep up. Rolling his chair away from his desk to sit across from me, occasionally jumping up to scribble something on his whiteboard, he explained synthetic biology to me by using as props the various objects cluttered on bookshelves and tables in his MIT office. To explain what MIT synthetic biologists meant by "abstraction," he held up a power cord connected to his laptop, unplugging it and then wrapping it around his hand several times. I don't need to understand what's happening inside this cable—the wiring, the circuitry inside the adapter, the current flowing from the outlet to the laptop—he told me, in order to get it to work. Why can't biology be similarly black-boxable, he asked, so that different levels of biological complexity (genetic sequences, genes, cells, tissues, organisms, ecosystems, etc.) are insulated from one another?

As the afternoon sun poured in through his office window, he next picked up four white plastic jars, labeled A, C, G, and T, handing me one of them. These jars held phosphoramidites—the raw materials of DNA synthesis. As he told me, these phosphoramidites, which were purified from either salmon milt or sugarcane, can be cheaply purchased online. This in itself was unsurprising. But then he offered a factoid intended to blow my mind: the raw materials I held in my hands contained enough nucleoside to synthesize the genome of every human on Earth.[3] He used these jars to demonstrate the second engineering principle to which MIT synthetic biologists ascribed: "decoupling." Simply, they believed that biological manufacture could and should be disconnected from biological design. Design work might be done on the computer, such as when a synthetic biologist contracts a DNA synthesis company to make an oligonucleotide (short genetic sequence) to order. Such bottles of phosphoramidite would then become raw material for the resulting sequence, and multiple oligonucleotides could be "glued" together to make larger genetic fragments, even whole genomes. In addition to separating design from manufacture, decoupling, I realized, was a desire to separate data from substance.

The person who drove efforts to "standardize" genetic material (the third engineering principle synthetic biologists hoped to install in their field) was Tom Knight, the principal investigator of the second laboratory at MIT in which I observed. One of Knight's most lasting contributions to synthetic biology was proposing in 2002 that genetic fragments could and should be "standardized," and indeed, that standardization was indispensable if synthetic biology were to succeed. Knight's thinking was shaped by his computer science training: he developed a standard for genetic "parts" explicitly modeled on the standard that circuit designers adopted for integrated-circuit design in 1980. Knight's standard stipulated that each genetic sequence should be cradled by a specific set of upstream and downstream restriction sites. His hope was that as ever more users borrowed and shared his standardized parts for use as components in engineered living systems, they would iteratively become something of an industry standard. In 2003 Knight published a full draft of his biological standard, naming the resulting entities "BioBricks." From Knight's initial infusion of 12 parts, over the next decade the number of standardized biological parts hosted by, stored at, and distributed by MIT snowballed to over 5,100 parts.

These two researchers, and the three engineering standards they espoused, directed the ambitions and purpose of graduate students in the MIT Synthetic Biology Working Group from 2005 to 2008.[4] During this time, I attended weekly laboratory meetings of the MIT Synthetic Biology Working Group, smaller group meetings devoted to specific research and publishing issues, and frequent public lectures, and I met informally with members of the Synthetic Biology Working Group. As a graduate student working with other graduate students, participating in the daily life of synthetic biologists entailed more than visiting labs and attending meetings and conferences—I also learned much over coffee and lunch, during cigarette breaks and parties. I spoke with numerous synthetic biologists and formally interviewed many of them. I also completed graduate courses in MIT's Department of Biological Engineering and observed undergraduate laboratories in the same department. E-mails, LISTSERVs, and wikis were forums enabling lively conversation. I participated in MIT's Synthetic Genomics Initiative, which was organized around discussions about the social and ethical implications of synthetic genomics.

Reading articles in scientific journals and attending journal clubs, I struggled to learn the language spoken by synthetic biologists and discovered that I was not alone in this effort. Because synthetic biology is a hybrid field, synthetic biologists trained in biology often had as hard a time understanding engineering terminology as those from engineering backgrounds had speaking biology. Over several years, I gradually grew proficient in this new technoscientific pidgin, able to discuss abstraction hierarchies in the same breath as zinc fingers. I tracked down mentions of synthetic biology in the popular press, noting each year how much harder it was becoming to stay abreast of that literature, as synthetic biology mushroomed from a minor movement centered at MIT into a more pervasive phenomenon.[5]

I attended annual professional conferences for synthetic biologists, often hosted in cities that already supported sizable technoscience research hubs (Zurich, Hong Kong). Such events were valuable because they allowed me a bird's-eye view of what synthetic biology looked like in any given year. As a dense meeting place for synthetic biologists of all kinds, conferences gave me ample opportunity to meet and interview a wide variety of researchers working in the United States, Europe, and Asia who had vastly divergent research agendas and were at different stages in their careers, from undergraduates to elder statespersons in the field.

Listening to lectures at an international conference in 2009, I was alarmed and dismayed, jotting in my field notes that most of the lectures bore nothing in common with what MIT synthetic biologists told me defined their field. It then occurred to me that my knee-jerk cynical reading (namely, that calling oneself a "synthetic biologist" attracted funding, and that therefore the field had become fragmented and disunified) was a gross oversimplification, and that this intellectual and practical diversity was in itself worthy of attention because the field had become more capacious than I could have known while embedded at MIT. Kristala Prather, an MIT synthetic biologist who hails from chemical engineering, joked that year that "if you ask five people to define synthetic biology, you will get six answers."[6] I realized that, if I were to understand what the field was becoming, the scope of my fieldwork would have to expand accordingly. The structure of *Synthetic* therefore mimics the growth of the field: as the field expanded away from MIT, so too did my fieldwork trek outward.

From 2009 to 2013 I followed synthetic biologists beyond the MIT Department of Biological Engineering and learned about other sorts of synthetic biology being practiced elsewhere, by people whose ideas about the field, its disciplinary precursors, and its aims were different from those I learned at MIT. Conducting fieldwork in the Bay Area in the summer of 2012, I reconnected with scholars I had met at earlier conferences, visiting laboratories ranging from academic (e.g., University of California, Berkeley) to federal (e.g., Lawrence Berkeley National Laboratory) to private corporations (e.g., Amyris Biotechnologies). The synthetic biologists whom I met at these labs, most of whom had not been reared at MIT, defined and practiced synthetic biology differently than did those I had worked with on the East Coast.

As synthetic biology grew in size, its definition expanded accordingly. The majority of synthetic biologists no longer embrace principles of standardization, abstraction, and decoupling (indeed, they were contentious even in the field's first years). Some synthetic biologists are frankly skeptical of what they perceive as MIT synthetic biologists aggressively evangelizing their standards, demanding that other researchers buy in to their model of how bioengineering should be done. In conversation, one synthetic biologist admitted to me that he was turned off by what he called "the East Coast vibe" in synthetic biology, which was "super all positive, group love," a façade he considered "bullshit," consisting of "dogma, shallow science, and a lot of paranoia."

The synthetic biologists I worked with on the West Coast were much more likely to have trained as chemical engineers, and their thinking hewed to comparisons of cells and microbes to tiny reactors in which chemical reactions cascaded. Yet, on both coasts, US synthetic biologists believe living substance to be mechanistic and partible; they expect it to behave no differently from inorganic matter. Today, the only remaining common denominator uniting most synthetic biological research is an avowedly engineering-based approach to manipulating living things.

In short, it is at times hard to distill that which unites the people and projects that travel under the name "synthetic biology." At the level of practice, the techniques, methods, and tools in synthetic biology labs are not unique—they overlap significantly with a wide variety of life science fields, including molecular biology and (in corporate spheres) biotechnology and pharmaceutical research. Synthetic biologists have radically different ideologies, especially when it comes to hot-button topics such as intellectual property. Their theories and practices are also widely divergent, even within a single laboratory. Synthetic biologists, by a pragmatic definition, are people who identify as synthetic biologists, who collaborate and work with other synthetic biologists, who attend synthetic biology conferences, or are attached to professional associations or institutions dedicated to synthetic biology research. Consequently, at a methodological level what unites this diverse cast of characters is *sociology*—simply put, I moved from one field site to another by following people.

Following the initial players in MIT synthetic biology, I sometimes found myself in unlikely destinations, ones that some (perhaps many?) synthetic biologists would not identify with "orthodox" synthetic biology: start-up companies, biohackers' private homes and community laboratories, workshops dedicated to reviving extinct species. These serendipitous termini, I believe, are one of the boons of anthropological fieldwork. Rather than treating the field as either homogeneous or unchanging, by following synthetic biology's earliest proponents I have had the opportunity to watch as a field grew, established itself, found footing in conferences and journals, agreed upon some norms of conduct, disagreed about much else, sprouted offshoots and subfields of its own, abandoned some founding principles, and developed technologies that rendered a sizable amount of synthetic biology research unrecognizable from what its founders had intended.

Acknowledgments

This book is *synthetic*. It is composed of a sprawling network of talented humans who plait the very marrow of this text. I can only hope that the resulting composition is viable. However, if it fails to thrive, then the fault is wholly my own.

First and foremost, I thank the synthetic biologists without whose magnanimity, wit, and long-suffering patience this book could not have been possible. I hope they will accept it as a necessarily inadequate gift in return for guiding me into their synthetic worlds. In particular, Drew Endy ushered me into the lively community of Cambridge synthetic biology, where I met Barry Canton, Austin Che, Caitlin Conboy, Heather Keller, Jason Kelly, Tom Knight, Sriram Kosuri, Natalie Kuldell, Alex Mallet, Julie Norville, Reshma Shetty, Pamela Silver, François St.-Pierre, Samantha Sutton, Ilya Sytchev, and Ty Thomson. Each of them has taught me how to think about life otherwise. Caroline Ajo-Franklin, Chris Anderson, Adam Arkin, Michelle Chang, George Church, Jay Keasling, Kristala Prather, and Chris Voigt welcomed me into their laboratories, where I tried my best not to get in the way. Many others, including Peter Ackermann, Greg Bokinsky, Rob Carlson, Kevin Costa, Mackenzie Cowell, Daniel Gibson, Rich Hansen, Jack Newman, Zach Serber, and Jeff Ubersax, were indulgent with both their time and their ideas, even when my questions tested the limits of their forbearance.

My colleagues at Harvard have made my academic home a vibrant and supportive one: Allan Brandt, Janet Browne, Jimena Canales, Alex Csiszar, Peter Galison, Jeremy Greene, Evelynn Hammonds, Anne Harrington, David Jones, Shigehisa Kuriyama, Rebecca Lemov, Naomi Oreskes, Katy Park, Ahmed Ragab, Sarah Richardson, Charles Rosenberg, Steven Shapin, and Heidi Voskuhl. Beyond my departmental orbit, I thank my interdisciplinary

interlocutors across campus for chance encounters and serendipitous dia-
logue at key moments: Robin Bernstein, Steph Burt, Jean and John Co-
maroff, Susan Greenhalgh, Nicholas Harkness, Sheila Jasanoff, Andrew Jewett, and Ajantha Subramaniam. Meredith Bircher, Sarah Champlin-Scharff,
Ellen Guarente, Paul Millett, Adekunle Ogunseye, Linda Schneider, Robin
Weston, and other Harvard administrators and staff are unflappable, having
come to my aid more times than I care to mention. Stefanie Chorianopoulos
and Brad Bolman were terrifically dedicated research fellows, whipping into
shape my haphazard databases.

   For reading and commenting on drafty chapters, sending me ethnographic tidbits, organizing panels, workshops, and seminars, nourishing
me with ongoing conversation (and not hesitating to argue with me), I
am grateful to a host of colleagues and friends who have helped me in
more ways than they know: Rachel Ablow, Julien Ayroles, Orkideh Beh-
rouzan, Etienne Benson, Mario Biagioli, Jeremy Blatter, Max Carradine,
Sean Clarke, Offir Dagan, Richard Doyle, Joseph Dumit, Michael Fischer,
Elaine Freedgood, Xiaolu Hsi, Daniel Jütte, Abhishek Kaicker, Christopher Kelty, Hannah Landecker, Vincent Lépinay, Lisa Messeri, Adam
Mestyan, Francesca Orsini, Nadya Peek, Laurie Penny, Ben Peter, Chris
Phillips, Hugh Raffles, Christian Reiss, Hans-Jörg Rheinberger, Astrid
Schrader, Hillel Schwartz, Susan Silbey, Alistair Sponsel, Hallam Stevens,
Emily Wanderer, Elizabeth Weed, Peter Weigele, Sharon Weltman, Ste-
ven Wilf, Rebecca Woods, Sara Wylie, and Laura and Jonathan Zittrain.
I have learned much from my teachers, more from my colleagues, and the
most from my students. Leah Aronowsky, Brad Bolman, Alyssa Botelho,
Richard Fadok, Kathryn Heintzman, Dani Inkpen, Abbas Jaffer, Colleen
Lanier-Christensen, Wythe Marschall, Eli Nelson, and Alexis Turner have
asked me exquisitely difficult questions and pushed me to clarify my thinking. Evelyn Fox Keller is my guru. She probably should be yours, too.

   Lucia Allais, Caroline Jones, Ewa Lajer-Burcharth, and Ruth Mack,
my Radcliffe writing group, provided a steady flow of red-inked margina-
lia, which they kindly buffered with an equal dose of encouragement. A
handful of big-hearted and brilliant people read my manuscript in full (or
very nearly so). In particular, Stefan Helmreich, David Jones, David Kai-
ser, Heather Paxson, and Harriet Ritvo were good enough to workshop
my manuscript (during summer break no less); their gimlet-eyed reading
and honest appraisal urged me through a final round of revisions. Jeremy
Greene and Rebecca Lemov—two-thirds of the Submariners—buoyed
me when I was submerged in prose and provided ballast when my thinking

was adrift. The feedback offered to me by Peter Galison, Emily Riehl, Charles Rosenberg, Steven Shapin, and Alma Steingart was thoughtful in every sense of the word and I am forever in their debt. Biogroop—composed of Stefan Helmreich, Rufus Helmreich, Natasha Myers, and Michael Rossi—is the brooding scholarly aggregate with whom I revel in writing against the grain of academic disciplinarity and propriety. The Exquisite Corpse has been a welcome respite between bouts of "serious" writing. This book would be much less without the incomparable delight of thinking with Biogroop these past ten years. The depths of Stefan Helmreich's intellect, generosity, and good humor are unfathomable. I would repay the debt I owe him if I could, but I fear his kindness has bankrupted me. At the very least, I aspire someday to be a mentor half as good as he.

I have presented portions of this book at a number of academic addresses—Brown University, Cornell University, the European University at Saint Petersburg, the University at Buffalo-SUNY, and York University. The incisive comments these audiences have provided have markedly improved this book. Eight years of fieldwork would not have been feasible without financial support from a National Science Foundation Doctoral Dissertation Research Improvement Grant (number 0847853), a National Science Foundation Graduate Research Fellowship, a Wenner-Gren Foundation Dissertation Fieldwork Grant (number 7928), as well as one blessedly quiet year spent as the Joy Foundation Fellow of the Radcliffe Institute for Advanced Study. At the University of Chicago Press, I thank Karen Darling, Pam Bruton, Susan Karani, Kristen Raddatz, and Evan White for their keen editorial eyes, boundless patience, and abiding faith in this work.

I thank my parents, Stuart and Andrea Roosth, and my sisters, Mira Roosth and Carolyn Agis, as well as my in-laws: Moshe Steingart, Ruth Jaffe, Daniela Steingart, and Neta Steingart. Laika reminds me how to achieve escape velocity from the drag of anthropocentrism. For showing me how we can make a life, and for perpetually being open to what that might mean, my deepest thanks are to my first and best reader, Alma Steingart.

Notes

Introduction

1. A description of my ethnographic methods may be found in the appendix.

2. RNA substitutes uracil (U) for thymine.

3. Ethnographer Susanna Finlay also noticed this: "Milling around the centre's laboratories providing assistance with electrophoresis gels, Polymerase Chain Reaction (PCR) runs and protein assays, to the untrained eye I could have been in any molecular biology laboratory." Finlay, "Engineering Biology?," 34.

4. Benner and Sismour, "Synthetic Biology," 533.

5. "Trends in Synthetic Biology Research Funding in the United States and Europe"; "Synthetic Biology, a New Paradigm for Biological Discovery."

6. In addition, the US synthetic biology terrain includes some 146 companies, 17 government laboratories, and 4 military laboratories. Synthetic Biology Project (2013), Woodrow Wilson International Center for Scholars, http://www.synbio project.org/library/inventories/map/.

7. "Trends in Synthetic Biology Research Funding in the United States and Europe."

8. Andrianantoandro et al., "Synthetic Biology"; Church, "Let Us Go Forth and Safely Multiply"; Forster and Church, "Synthetic Biology Projects in Vitro"; Purnick and Weiss, "Second Wave of Synthetic Biology."

9. Keasling's laboratories will be highlighted in chapters 2 and 4. Ro et al., "Production of the Antimalarial Drug Precursor Artemisinic Acid in Engineered Yeast."

10. Gibson et al., "Complete Chemical Synthesis, Assembly, and Cloning of a *Mycoplasma genitalium* Genome."

11. I further describe this research in chapter 3.

12. Wade, "Researchers Say They Created a 'Synthetic Cell.'" This organism was not, however, *de novo* life—much more than computers was necessary to "parent" this creature. The nucleotides used in synthesis no doubt were harvested

from sugarcane or salmon milt, portions of the genome were inserted into yeast cells for assembly, and the genome itself was hosted and replicated by a recipient bacterium.

13. Helmreich, "What Was Life?"; Helmreich, *Alien Ocean*.

14. Rob Carlson, MIT lecture, Cambridge, MA, November 7, 2008.

15. Jha, "From the Cells Up."

16. The sorts of disciplinary histories synthetic biologists told themselves about themselves are "usable pasts": "less technical and more accessible narratives that make sense of the contemporary world by reflecting on the past and its difference from today." Kelty, *Two Bits*, 65.

17. This account is further detailed in chapter 1.

18. Feferman, *Simple Rules . . . Complex Behavior*. See also Prusinkiewicz and Lindenmeyer, *Algorithmic Beauty of Plants*. Langton quoted in Helmreich, *Silicon Second Nature*, 90. For an account of how Langton's abductive thinking is consonant with a longer history of generalizing biological theories by divorcing logical forms from their physical counterparts, see Helmreich and Roosth, "Life Forms."

19. Feynman, "Feynman's Office."

20. Carr and Church, "Genome Engineering," 1152.

21. Sagmeister and Endy, "On Design," 71.

22. Dear, "What Is the History of Science the History Of?"; Shapin, "House of Experiment in Seventeenth-Century England"; Smith, *Body of the Artisan*.

23. Foucault, *Order of Things*, 175.

24. W. Coleman, *Biology in the Nineteenth Century*; Nyhart, *Modern Nature*; Nyhart, *Biology Takes Form*; Richards, *Romantic Conception of Life*.

25. Richards, *Meaning of Evolution*; Cunningham and Jardine, *Romanticism and the Sciences*; Lenoir, *Strategy of Life*; Rupke, *Richard Owen*; Richards, *Tragic Sense of Life*.

26. Browne, *Charles Darwin*, vols. 1–2; Mayr, *Evolutionary Synthesis;* Smocovitis, *Unifying Biology*; Numbers, *Darwinism Comes to America*; Bowler, *Non-Darwinian Revolution*; Kohler, *Landscapes and Labscapes*.

27. W. Coleman and Holmes, *Investigative Enterprise*; Holmes, "Old Martyr of Science"; Holmes, "Claude Bernard, the 'Milieu Intérieur,' and Regulatory Physiology"; Normandin, "Claude Bernard and *An Introduction to the Study of Experimental Medicine*"; Gross, "Claude Bernard and the Constancy of the Internal Environment"; Todes, *Pavlov's Physiology Factory*.

28. Brain, *Pulse of Modernism*; Cartwright, " 'Experiments of Destruction' "; Chadarevian, "Graphical Method and Discipline"; Lenoir, *Inscribing Science*.

29. Landecker, "New Times for Biology"; Landecker, *Culturing Life*.

30. Pauly, *Controlling Life*, 7–8.

31. Creager, *Life of a Virus*; Kohler, *Lords of the Fly*; Landecker, *Culturing Life*; Rader, *Making Mice*.

32. Kay, *Molecular Vision of Life*; Kay, *Who Wrote the Book of Life?*; Keller,

Making Sense of Life; Chadarevian, *Designs for Life*; Creager, *Life of a Virus*; Rheinberger, *Toward a History of Epistemic Things*.

33. The aphorism was first published in Monod and Jacob, "General Conclusions," 393. Monod, however, first dates it to a lecture he gave in New York in 1954. For a more detailed account of Monod's usage, as well as similar antecedents claiming biomolecular unity, see Friedmann, "From *Butyribacterium* to *E. coli*."

34. Keller, *Secrets of Life, Secrets of Death*.

35. Bernal, "Definitions of Life," 13; quoted in Keller, *Secrets of Life, Secrets of Death*, 96.

36. Fortun, "Human Genome Project and the Acceleration of Biotechnology"; Kevles, *Code of Codes*.

37. H. Stevens, *Life out of Sequence*.

38. Anderson, "End of Theory."

39. Now is not the first time that scientists or social observers have cautioned that science is in extremis. One of the most recent doomsayers was mathematician John Horgan, who in *The End of Science* diagnosed science as entering a period of "diminishing returns," in which the great scientific problems have been solved and all that is left is increasingly complex and detail-oriented problem solving that will fail to inspire a new generation of scientists, whom he already worried might be wooed by the siren song of the liberal arts and illicit substances. Situating himself squarely in the United States in the 1990s, Horgan further worried that "postmodern" approaches like simulation risked undermining empirical science and its claims to "Truth." Perhaps every generation of scientists sees science endangered by some threat, whether it be complexity, decreased funding, increased regulation, or more obscure specters (e.g., digital simulation, a paucity of interesting problems, or the "social construction" debates of the 1990s Science Wars). Ross, *Science Wars*; Ashman, *After the Science Wars*; Labinger and Collins, *One Culture?*

40. Further, "a history of data depends on an understanding of the material culture—the tools and technologies used to collect, store, and analyze data—that makes data-driven science possible." So argue historians of science engaged in an ongoing and collaborative project, Historicizing Big Data, at the Max Planck Institute for the History of Science. See http://www.mpiwg-berlin.mpg.de/en/research/projects/DeptII_Aronova_Oertzen_Sepkoski_Historicizing.

41. Strasser, "Genbank"; Strasser, "Experimenter's Museum"; Strasser, "Collecting Nature."

42. To challenge your knowledge of contemporary -omics, try your hand at a crossword puzzle published in *Nature*: M. Baker, "Big Biology."

43. Galison, *Einstein's Clocks, Poincaré's Maps*.

44. Spivak, "Subaltern Studies," 10.

45. Clifford and Marcus, *Writing Culture*; Marcus and Fischer, *Anthropology as Cultural Critique*.

46. For an account of religious language infusing Artificial Life research, see Helmreich, "The Spiritual in Artificial Life."

47. Spivak, "Subaltern Studies."

48. Jane Calvert argues similarly, and in a viral idiom: "The standardized, modular, decomplexified creations of synthetic biology will inevitably start to infect our understandings of what is 'natural,' which . . . is itself a 'receding horizon' defined primarily in terms of what it is opposed to." Calvert, "Synthetic Biology," 108.

49. On "reparative reading," see Sedgwick, "Paranoid Reading and Reparative Reading."

Interlude One

1. The dictionary definitions introducing each interlude are inspired by and loosely based on the *Oxford English Dictionary*, first edition (1919), second edition (1989), and online editions (2000 and 2016).

2. J. Schneider, "In and out of Polyester," 6.

3. Handley, *Nylon*, 28.

4. Kanigel, *Faux Real*.

5. Schnapp, "Fabric of Modern Times," 210.

6. Ibid., 195-96.

7. Though the Italian futurists and their passion for speed, violence, and the synthetic aesthetic may seem far afield from synthetic biology, Drew Endy draws an explicit comparison between the two movements. When in a 2006 laboratory meeting I was asked to report on my research progress, I idly commented that I wanted to find a point of comparison to synthetic biology, assuming that the assembled researchers might suggest another scientific field in which I might do further ethnographic work. Endy surprised me by immediately suggesting that a logical comparison would be Italian Futurism, as both movements seek to destroy the status quo and embrace the future as an aesthetic good.

8. Blaszczyk, "Styling Synthetics," 486.

9. T. Taylor, *Strange Sounds*, 76.

10. DeLillo, *White Noise*, 52.

11. See http://www2.dupont.com/Phoenix_Heritage/en_US/1915_c_detail.html.

12. Handley, *Nylon*; Ndiaye, *Nylon and Bombs*.

13. Altman, "Michael DeBakey, Rebuilder of Hearts, Dies at 99."

14. See http://www2.dupont.com/Phoenix_Heritage/en_US/1915_c_detail.html.

15. Blaszczyk, "Styling Synthetics," 489.

16. Ibid., 493.

17. Amato, *Stuff*; Bensaude-Vincent, "Construction of a Discipline"; Hounshell and Smith, *Science and Corporate Strategy*.

18. Carson, *Silent Spring*, 7.

19. J. Schneider, "In and out of Polyester."

20. Ibid., 2.

21. Compiling a list of terms that might occupy shared valences with synthetic biology, then, we might also count "Artificial Life," a field that was for a time also referred to as "synthetic biology." See interlude 6, SYNTHETIC *adj. 6.*

22. The trend for synthetic foods is marked by the appearance of "synthetic cream" in the second edition of the *Dictionary of Dairying* in 1955.

23. Belasco, "Algae Burgers for a Hungry World?," 608; Belasco, *Meals to Come*; Belasco, "Future Notes."

24. Frohlich, "Accounting for Taste."

25. In the present day, synthetic biology's promise to develop products that will meet and solve similar issues (e.g., biofuels to meet gasoline shortages, antimalarial drugs to staunch "orphan diseases" in overpopulated countries) resonates with these earlier uses of "synthetic" to denote putative magic bullets to anthropogenic troubles.

26. I oppose this usage to earlier meanings of "synthetic" to refer to either a mode of thinking or a scientific method. See interlude 5, SYNTHETIC *adj. 5.*

27. Harvard synthetic biologist George Church compares synthetic biology to materials science and the synthetic materials industry. In an e-mail, he identified five strains of synthetic biology, one of which he notated "PL," for "plastic life." He described this philosophy as being "[i]nspired by life, but using basic components not found in life—analogous to nylon being 'synthetic silk.' Branches of PL include replicating robots, 'Alife' (= Artificial Life) in computers, and in silico biology." Correspondence with author, June 2012.

Chapter One

1. Tolerated, at least, among male faculty members, and especially those in natural science and engineering departments.

2. Once, upon bumping into Endy in the hallway outside the laboratory, I was surprised to find him dressed in a button-down shirt. Taken aback, I asked why he was so formally dressed. He glumly explained that he had met with MIT alumni and investors that morning and had been advised by colleagues to dress up for the meeting.

3. Pennisi, "Synthetic Biology Remakes Small Genomes," 769.

4. Once a phage has infected a cell, it can replicate in one of two ways: the lytic cycle or the lysogenic cycle. Lysis means that the phage uses the host cell to produce more viral nucleic acid, forming new virus that then burst the cell. Lysogenesis means that viral nucleic acid embeds in that of its host and is transmitted to daughter cells as the cell divides. Rejecting the possibility that the switch between lysogenesis and lysis is random, Endy paraphrased Einstein to his graduate

students: phage "doesn't play dice." When another graduate student published in the *Proceedings of the National Academy of the Sciences*, grad student Jason Kelly's congratulatory e-mail announcement to the laboratory read "Does lambda [phage] play dice?"

5. This passion for children's construction toys is reflected in the way synthetic biologists talk about the genetic fragments they seek to standardize, which they overwhelmingly compare to Legos. Colin Milburn argues that comparisons of atoms to Tinkertoys in nanotechnology are *"tropic protocols:* figurative language games that actually script out certain technical, epistemic, and social dimensions of laboratory research." Milburn, "Just for Fun," 230. As such, the "ludic and the ludicrous have been inconspicuously modulating not only the social dynamics of the molecular sciences, but also the very content of those sciences" (231–32). Figuring genetic sequences as Legos shapes researchers' thinking about synthetic biology, turning it into an extended game, one that is playful, innocent, and innocuous.

6. Holt, "History Keeps Bethlehem Steel from Going off the Rails."

7. Bahnisch, "Embodied Work, Divided Labour"; Stewart, "Management Myth"; F. Taylor, *Principles of Scientific Management.*

8. The Taylorization of industrial synthetic biology is the subject of chapter 4.

9. Holmes and Summers, *Reconceiving the Gene.*

10. Putnam et al., "Biochemical Studies of Virus Reproduction."

11. Volkin, Astrachan, and Countryman, "Metabolism of RNA Phosphorus in *Escherichia coli* Infected with Bacteriophage T7," 554.

12. Dunn, Studier, and Gottesman, "Complete Nucleotide Sequence of Bacteriophage T7 DNA and the Locations of T7 Genetic Elements."

13. Endy, "Development and Application of a Genetically-Structured Simulation for Bacteriophage T7," 1.

14. Daston and Galison, *Objectivity.*

15. Atwood, *Oryx and Crake*, 206.

16. Chadarevian and Hopwood, *Models*; Francoeur, "Forgotten Tool"; Hesse, *Models and Analogies in Science*; Klein, *Tools and Modes of Representation in the Laboratory Sciences*; Klein, *Experiments, Models, Paper Tools*; Morgan and Morrison, *Models as Mediators*; Sismondo, "Models, Simulations, and Their Objects"; Waldby, *Visible Human Project.*

17. Kevles, *Code of Codes.* Phenotype refers to the set of observable characteristics that result from an organism's genetics being regulated and modified by its environment. For a critique of the division between genetics and environment, or "nature" and "nurture," see Keller, *Mirage of a Space between Nature and Nurture.*

18. Davis, "Human Genome and Other Initiatives," 343.

19. Roberts, "Tough Times Ahead for the Genome Project."

20. Davis, "Human Genome and Other Initiatives," 343.

21. F. Collins, "Has the Revolution Arrived?"; Venter, "Multiple Personal Genomes Await."

22. Clinton, "Reading the Book of Life."

23. R. Carlson, "Pace and Proliferation of Biological Technologies"; Mardis, "A Decade's Perspective on DNA Sequencing Technology."

24. Textual metaphors for genetic material and their impact on thinking about property rights in synthetic biology are addressed in chapter 3. For analyses of the DNA-as-code metaphor, consult Doyle, *On Beyond Living*; Haraway, *Modest_Witness@Second_Millennium.FemaleMan_Meets_OncoMouse*; Kay, *Who Wrote the Book of Life?*; Keller, *Refiguring Life*; R. Mitchell and Thurtle, *Data Made Flesh*.

25. Quote in Doyle, *On Beyond Living*, 14.

26. TMSI was originally named the Philip Morris Molecular Sciences Institute, although after its name and funding source caused a media furor, Brenner changed the name. Philip Morris wanted researchers at the institute to study the molecular basis for lung cancer under the broad canopy of research into cell signal transduction. In its first year, TMSI scientists published papers suggesting that smoking lowered the risk of Alzheimer's disease. For historical accounts of how tobacco funding has inflected scientific research, see Brandt, *Cigarette Century*; Oreskes and Conway, *Merchants of Doubt*; Proctor, *Cancer Wars*; Proctor, *Golden Holocaust*.

27. Endy and Brent, "Modelling Cellular Behaviour," 394.

28. Yet Endy and Brent predict that biological modeling would nonetheless be a welcome supplement to the social mechanisms by which experimental knowledge is currently vetted: "Biology currently tests the validity of qualitative conclusions from different laboratories by mechanisms that range from peer-review to gossip. These are fairly effective." Nonetheless, computational models would, they hoped, be better than social evaluation of experimental evidence because "providing access to simulations to large communities of biologists should accelerate the process of biological discovery itself" (ibid., 395). As synthetic biology would later develop at MIT, researchers put in place Open Source approaches to standardizing and speeding up research collaboration by building an online infrastructure halfway between "peer-review" and "gossip," a platform they named OpenWetWare.

29. Chan, Kosuri, and Endy, "Refactoring Bacteriophage T7," E1.

30. Ibid., E2.

31. Lee, "Meaning of Molluscs," 227.

32. Galison, "Aufbau/Bauhaus."

33. Pehnt, "Gropius the Romantic."

34. In *Silicon Second Nature*, Stefan Helmreich identifies the heterosexual, Judeo-Christian, sociobiological, and bureaucratic assumptions built into Koza's genetic programming.

35. Sussman, like Endy, promoted free software approaches to intellectual property. See chapter 3.

36. Chan, Kosuri, and Endy, "Refactoring Bacteriophage T7," E1 (emphasis added).

37. Ibid.

38. The technological specimens (such as electronic tubes) arranged by Gilbert Simondon in 1958 reverse this thinking, comparing design to evolution and machines to living things (rather than vice versa). Simondon suggests that design is itself an evolutionary process, a claim that minimizes the input of actual engineers and designers in the gradual modification of motors and telephones, even as it renders technological objects at least analogously "lifelike." Simondon, "Genesis of the Individual."

39. Keller, "Organisms, Machines, and Thunderstorms, Part One," 50–51 (emphasis in original).

40. Quoted in Bowler, "Darwinism and the Argument from Design," 29.

41. Paley, *Natural Theology*, 6.

42. Riskin, "Divine Optician," 362.

43. Ibid., 352.

44. Ospovat, "God and Natural Selection," 187. Also quoted in Bardini, *Junkware*, 57 (emphasis in original).

45. Bardini, *Junkware*, 59.

46. Quoted in Keller, "Organisms, Machines, and Thunderstorms, Part Two," 19.

47. This fact adds a new dimension to Jane Calvert's argument that "[b]oth engineering and design are activities that incorporate broader social goals and values." Sometimes, social goals direct not only how a living thing is designed but also how it becomes part of the milieu by reference to which its fitness is evaluated. Calvert, "Engineering Biology and Society," 417.

48. The notion that synthetic biology might salvage life from bad design or junky programming was not limited to MIT's synthetic biologists. For example, Alex Sunguroff, at the time president of a Massachusetts-based start-up company that sold in vitro platforms that mimicked "biologically realistic microenvironments," commented: "It might be true that to understand, best, the coding of life, we must study the computer code generated by incompetent programmers. Insights from examples of how not to code, might best illuminate the techniques that evolved to process information in life." Elaborating on this description, Sunguroff continued, "the fact that the guiding principles of life are random evolution with immediate expediency, implemented in a wetware environment, results in code that is poor in subroutinizing and overloaded in global variables . . . it's still so fudgey [*sic*]" (e-mail to Synthetic Biology LISTSERV, May 18, 2006). Menaced by misinformation, bad information, or noise, life is, in this depiction, best understood by studying incompetent programmers.

49. Magnetotactic bacteria orient themselves along magnetic field lines.

50. Personal communication, April 18, 2006 (emphasis added).

51. Numbers, *Darwinism Comes to America*.

52. Austin Che, e-mail to Synthetic Biology LISTSERV, December 20, 2005.

53. "*Nature* Cover Exploits Intelligent Design while Inside Attacks It," Novem-

ber 24, 2005, *Creation-Evolution Headlines*, ed. David Coppedge, http://creation safaris.com/crev200511.htm.

54. Sprinzak and Elowitz, "Reconstruction of Genetic Circuits," 447.

55. Siegel, "Interview: Drew Endy," 32.

56. Philosopher of science Elliott Sober ends his essay on the late twentieth-century ID controversy by predicting that bioengineering will not foreclose arguments in favor of ID. Rather, bioengineering will make such debates practical, rather than theological, questions: can you spot which is the intelligently designed organism? When humans "build organisms from nonliving materials," he writes, it "will not close down the question of whether the organisms we observe were created by intelligent design or by mindless natural processes; on the contrary, it will give that question a practical meaning, since the organisms we will see around us will be of both kinds." Sober, "Design Argument," 122.

57. Latour, "A Cautious Prometheus?," 2.

58. Ibid., 11.

59. Ibid., 5.

60. "Meanings of 'Life,'" 1031.

61. Indeed, by 2010 one of the recommendations delivered in a report on synthetic biology by the Presidential Commission for the Study of Bioethical Issues addressed this very concern: "the use of sensationalist buzzwords and phrases such as 'creating life' or 'playing God' may initially increase attention to the underlying science and its implications for society, but ultimately such words impede ongoing understanding of both the scientific and ethical issues at the core of public debates on these topics." Gutmann and Wagner, *New Directions*, 15.

62. Despite multiple efforts to conduct fieldwork in JCVI or interview Craig Venter, I was never able to do so. As a result, my descriptions of Venter are necessarily limited to popular press and the ways in which MIT synthetic biologists explained and responded to his work.

63. Gibbs, "Creation Myths," 60.

64. For an example, see Highfield, "From Reading DNA to Writing It."

65. Moses, "Intelligent Design."

66. While hubris is here a clear theme, not all synthetic biologists were opposed to excessive pride. One graduate student explained her thinking on the aim of synthetic biology: "Should we not dream of trying things that have never been done, of using engineering skill to create the new? Process engineering is boring. Instead, let us create new things to improve the world. There are so many beautiful things to do. Icarus burned, but there was beauty in his burning."

67. Gen. 3:3–4 (Authorized, or King James, Version).

68. Similar Edenic imagery has appeared in Artificial Life research. See Helm-reich, *Silicon Second Nature*.

69. Ibid., 201.

70. Church, "Constructive Biology."

71. Ibid. (emphasis added).

72. Ibid.

73. Doyle, *Darwin's Pharmacy*.

74. Ansell-Pearson, *Viroid Life*; Bardini, *Junkware*; Helmreich, "Flexible Infections"; MacPhail, "Viral Gene."

Interlude Two

1. Barr, *Cubism and Abstract Art*, quoted in Robbins, "Abbreviated Historiography of Cubism," 280.

2. Cottington, *Cubism and Its Histories*; Barr, *Cubism and Abstract Art*.

3. Gris, *Bulletin de la vie artistique*, quoted in Robbins, "Abbreviated Historiography of Cubism," 281.

4. Kahnweiler, *Rise of Cubism*, 12.

5. Marinetti, Settimelli, and Corra, "Synthetic Futurist Theatre," 134 (emphasis in original).

6. Higgins, "Fluxus," quoted in Milman, "Futurism as a Submerged Paradigm for Artistic Activism and Practical Anarchism," 162.

7. Leach, *Revolutionary Theatre*, 120.

8. Lopukhov, *Writings on Ballet and Music*, 15. See also Salter, *Entangled*; Torda, "Tairov's 'Princess Brambilla.'"

9. Robbins, "Abbreviated Historiography of Cubism," 282.

Chapter Two

1. Woese, Kandler, and Wheelis, "Towards a Natural System of Organisms."

2. In *Synthetic Aesthetics*, Ginsberg collaborates with synthetic biologists and sociologists studying synthetic biology to interrogate what it means to "design nature." In her *E. chromi* project, she worked with an International Genetically Engineered Machine team at Cambridge, using BioBrick parts to engineer bacteria that express colors: red, yellow, green, blue, violet, and brown. She then imagined a future in which such bacteria could be used as indicators by changing color in the presence of certain toxins or diseases.

3. *Nucleic Acids Research* 38, no. 8 (2010): i; Baldwin, *Synthetic Biology*.

4. Linné, *Systema Naturae*.

5. Haeckel, *Generelle Morphologie der Organismen*.

6. Chatton, *Titres et travaux scientifiques (1906–1937) de Edouard Chatton*.

7. Copeland, *Classification of Lower Organisms*.

8. Whittaker, "New Concepts of Kingdoms of Organisms."

9. Woese, Kandler, and Wheelis, "Towards a Natural System of Organisms."

10. Ginsberg, "Synthetic Kingdom." Ginsberg's invocation of the first-person plural is problematic, marking herself as both coextensive with synthetic biologists and enrolling all people in an unmarked "we" into the project of synthetic biology.

11. For an excellent introduction to 250 years of naturalism and classification, consult Farber, *Finding Order in Nature*.

12. Daston and Park, *Wonders and the Order of Nature*.

13. Endersby, *Imperial Nature*; Findlen, *Possessing Nature*; Jardine, Secord, and Spary, *Cultures of Natural History*; Ritvo, *The Platypus and the Mermaid*; Spary, *Utopia's Garden*.

14. Koerner, *Linnaeus*.

15. Hagen, "Development of Experimental Methods in Plant Taxonomy"; Hagen, "Experimentalists and Naturalists in Twentieth-Century Botany"; Hagen, "Ecologists and Taxonomists"; Hagen, "Retelling Experiments"; Hagen, "Naturalists, Molecular Biologists, and the Challenges of Molecular Evolution"; Hagen, "Origins of Bioinformatics"; Strasser, "Laboratories, Museums, and the Comparative Perspective."

16. For example, the "hacked" phylum is distinguished from the "advanced genetically engineered" one. This classificatory principle is further emphasized in Ginsberg's artist statement, in which she distinguishes between "DIY hacked" bacteria and "entirely artificial corporate life-forms." A second classificatory principle is also discernible in this tree of life. Synthetic living systems, it suggests, will come to be classified according to their features and functions—not by what they *are* but by what they can *do*: "engineered life will compute, produce energy, clean up pollution, make self-healing materials, kill pathogens and even do housework." Ginsberg, "Synthetic Kingdom."

17. Franklin, *Dolly Mixtures*; Haraway, *Modest_Witness@Second_Millennium. FemaleMan_Meets_OncoMouse*.

18. Haraway, *Modest_Witness@Second_Millennium.FemaleMan_Meets_Onco-Mouse*, 56.

19. Loeb, *Organism as a Whole*, 319.

20. Owen is a pseudonym.

21. Peter Ackermann, interview with the author, August 13, 2012.

22. Anonymous informant in interview with the author, 2013.

23. For one critique of the de-queering of LGBTQ politics, see Warner, *Trouble with Normal*. For a refusal of queer reproduction, consult Edelman, *No Future*.

24. D. Schneider, *American Kinship*; Rubin, "Traffic in Women."

25. Rubin, "Traffic in Women," 174.

26. In asking how lesbian couples undermine such economic transactions, Rubin anticipates Luce Irigaray question, "But what if the 'goods' refused to go to market? What if they maintained among themselves 'another' kind of trade?" Irigaray, *Speculum of the Other Woman*, 110.

27. Cultural anthropologist Marilyn Strathern instead points out that nature

and culture are not universal categories but instead are Euro-American analytics that anthropologists sometimes apply to cultures with orthogonal conceptualizations: "these two domains are not brought into systematic relationship; the intervening metaphor of culture's dominion over nature is not there." Strathern, "No Nature, No Culture."

28. Tyson, "What's the Next Big Thing?"

29. Service, "Rethinking Mother Nature's Choices," 793.

30. JBEI partners the Lawrence Berkeley National Laboratory, Sandia National Laboratories, UC Berkeley, UC Davis, the Carnegie Institution for Science, and the Lawrence Livermore National Laboratory.

31. Hsu, "Reflections on the 'Discovery' of the Antimalarial Qinghao."

32. Ro et al., "Production of the Antimalarial Drug Precursor Artemisinic Acid in Engineered Yeast," 940.

33. T. Mitchell, *Rule of Experts*, 22, 27.

34. Ibid., 52–53.

35. World Health Organization, *World Malaria Report*. Later press releases predicted that the dose, once on the market, would cost closer to fifty US cents.

36. The biofuels project at Amyris Biotechnologies is the subject of chapter 4.

37. Martin et al., "Synthetic Metabolism," 277.

38. Ibid., 283.

39. Ibid., 282.

40. Tawfik, "Messy Biology and the Origins of Evolutionary Innovations," 692 (emphasis added).

41. "Unnatural" is their term, not mine. In published literature, it is noteworthy that the technical term for the hybrid genetic systems that synthetic biologists cobble together is "*unnatural*" biocatalytic pathways.

42. Weston, "Forever Is a Long Time."

43. See, e.g., Franklin, "Biologization Revisited," 303.

44. Jackie Stacey adopts the work of Kath Weston to examine the queer kinships represented in science-fictional and horror films about genetic engineering, suggesting that such kinships displace blood and genealogy in favor of "alternative bonds of relatedness and forms of intimacy." Stacey, *Cinematic Life of the Gene*, 117. On transgenics and relatedness, see also Haraway, *Modest_Witness@Second_Millennium.FemaleMan_Meets_OncoMouse*.

45. Franklin, "Biologization Revisited," 302.

46. Ibid. (emphasis in original), 314.

47. Hayden, "Gender, Genetics, and Generation," 44. See also Weston, *Families We Choose*.

48. Emma Frow and Jane Calvert argue that iGEM enrolls undergraduate students into a "moral economy" of shared parts: Frow and Calvert, " 'Can Simple Biological Systems Be Built from Standardized Interchangeable Parts?' " For a sociological account of iGEM, see Balmer and Bulpin, "Left to Their Own Devices."

49. I return to the ways in which species boundaries are rendered problematic by whole-genome synthesis in chapter 6.

50. Douglas, *Purity and Danger*, 69.

51. It is important to note here that Arkin argues from principles he has learned in molecular biology, not rabbinic law. The former Sephardi chief rabbi of Israel came to a similar conclusion, although by a different route, arguing that genes from nonkosher animals do not affect the *kashrut* of a recipient species because genes, he claims, are not considered "food" and have no taste. Further, he decides, they do not physically or behaviorally modify fish or mammals (e.g., split hooves, presence or absence of fins or scales) or transmute or ferment raw substances (as in cheese and wine) in a manner that would impact their *halachic* status. Bakshi-Doron, *Binyan Av* (my translation).

52. Delebecque, "Bacteria, Archaea, Eukarya ... + Synthetica?" (my emphasis).

53. Craig Venter forwarded a radically different model of synthetic parentage when he pronounced that the minimal form of life later nicknamed "Synthia" (which J. Craig Venter Institute researchers manufactured by synthesizing the genome of *Mycoplasma mycoides* and transplanting it into a *Mycoplasma capricolum* cell) was "the first living self-replicating species to have a computer as a parent." Venter, *Life at the Speed of Light*, 125. See also Roosth, "Godfather, Part II." His assertion calls to mind media theorist N. Katherine Hayles's *My Mother Was a Computer*, a title that Hayles imagines to be the response an "artificial-life simulation might give if asked who its parent was," even as the title imagines that computers might already efface or replace "Mother Nature." Hayles, *My Mother Was a Computer*, 5. If Venter is to be believed, Synthia is the lawful offspring of both him and a computer *genetrix*.

54. Butler, *Bodies That Matter*, 95.

55. Weston, *Families We Choose*, 40 (emphasis in original).

56. Bernstein and Reimann, *Queer Families, Queer Politics*.

57. Hayden, "Gender, Genetics, and Generation," 41.

58. Hoefinger, " 'Professional Girlfriends,' " 252–53.

59. Strathern, review of *Families We Choose*, 196.

60. Weston's analysis has since been critiqued for being overly limited to white, middle- and upper-class, liberal urban queers, whose utopian affinal bonds of "choice" might not be afforded by less affluent Americans. So too, synthetic biologists' ideological queering of biological relatedness is also a utopian endeavor embedded in particular relations of power, liberal politics, and economic potentials.

61. Hayden, "Gender, Genetics, and Generation," 45.

62. Helmreich, *Alien Ocean*, 23. See also Helmreich, "Nature/Culture/Seawater."

63. In this respect, my project sympathizes with anthropologists who suggest that Melanesian anthropology, and in particular Melanesian kinship, can help us to think about biology and biotechnology otherwise. Bamford, *Biology Unmoored*; Strathern, *Reproducing the Future*; Strathern, *Kinship, Law, and the Unexpected*.

64. Such an approach has elsewhere been described as "lateral thinking": Maurer, *Mutual Life, Limited*; Hansen, "Adapting in the Knowledge Economy."

65. Hayden, "Kinship Theory, Property, and the Politics of Inclusion," 342.

66. Ibid., 339 (emphasis in original).

67. Haraway, *Modest_Witness@Second_Millennium.FemaleMan_Meets_OncoMouse*, 53.

68. "In the processes of materialized refiguration of the kinship between different orders of life, the generative splicing of synthetic DNA and money produces promising transgenic fruit. Specifically, natural kin becomes brand or trademark, a sign protecting intellectual property claims in business transactions." Ibid., 65–66.

69. Hayden, "Kinship Theory, Property, and the Politics of Inclusion," 340.

Interlude Three

1. Bensaude-Vincent, "Synthetic Biology as a Replica of Synthetic Chemistry?," 314.

2. Bensaude-Vincent and Stengers, *History of Chemistry*, 144.

3. Van den Belt and Rip, "Nelson-Winter-Dosi Model and Synthetic Dye Chemistry," 145.

4. Bensaude-Vincent and Stengers, *History of Chemistry*, 145.

5. Liebig, *Lettres sur la chimie*, quoted in Bensaude-Vincent and Stengers, *History of Chemistry*, 146.

6. Leslie, *Synthetic Worlds*, 78.

7. Pickstone, "Sketching Together the Modern Histories of Science, Technology, and Medicine," 130.

8. Yeh and Lim, "Synthetic Biology," 522, 523.

9. Bensaude-Vincent, "Synthetic Biology as a Replica of Synthetic Chemistry?," 316.

Chapter Three

1. Think of Cold War eschatological paranoia and salvational hopes, the privileging of genetic form as transcendent above bodily substance, reconfigured creation narratives told and retold in biotechnology and Artificial Life research, and DNA as a sort of sacred latter-day soul, bearing either original sin or predestiny. Kay, *Who Wrote the Book of Life?*; Nelkin, "Genetics, God, and Sacred DNA."

2. Kay, *Who Wrote the Book of Life?*, 14.

3. Clinton, "Reading the Book of Life."

4. Carroll, Donnelly, and O'Sullivan, "'We Are Learning Language in Which God Created Life.'"

5. Meek and Ellison, "On the Path of Biology's Holy Grail."

6. Private lecture, Harvard University, November 2005.

7. Endy, Bassler, and Carlson, *Overview and Context of the Science and Technology of Synthetic Biology*, 6.

8. Johns, *Nature of the Book*, 1.

9. This method is indebted to, among others, ibid.; Johns, *Piracy*; Biagioli, "Between Knowledge and Technology."

10. *Eau d'E. coli* was a synthetic bacterium engineered by undergraduates at MIT competing in the 2006 International Genetically Engineered Machine competition. The system expressed phase-dependent odors, smelling like either bananas or wintergreen at different stages of its growth. PoPS was a standard measurement inaugurated by Caitlin Conboy of the MIT Synthetic Biology Working Group. "Polymerase per second" is a metric for evaluating the rate of transcription by measuring how many RNA polymerase molecules pass a specified section of DNA every second.

11. Calvert, "Commodification of Emergence," 384.

12. Rose, "Mothers and Authors," 614.

13. Patent Act of 1793, Ch. 11, 1 Stat. 318–23 (February 21, 1793).

14. Hotchkiss v. Greenwood, 52 U.S. 11 How. 248 (1850).

15. *Ex Parte Latimer*, CD, 46 OG 1638, *US Patent Office, Decisions of the Commissioner of Patents and of the United States Courts in Patent Cases* (1889): 123–27.

16. Bugos and Kevles, "Plants as Intellectual Property"; Kevles, "Patents, Protections, and Privileges"; Kevles, "Protections, Privileges, and Patents"; Kloppenburg, *First the Seed*.

17. Diamond v. Chakrabarty, 447 U.S. 303, 100 S. Ct. 2204, 65 L. Ed. 2d 144 (1980). See also Hughes, "Making Dollars out of DNA"; Kevles, "Ananda Chakrabarty Wins a Patent."

18. Hughes, *Genentech*; Wright, "Recombinant DNA Technology and Its Social Transformation."

19. Greenberg and Kamin, "Property Rights and Payment to Patients for Cell Lines Derived from Human Tissues"; Landecker, "Between Beneficence and Chattel."

20. Kevles, "Of Mice and Money."

21. See chapter 1.

22. The Bayh-Dole Act, which passed the same year, effectively joined academic research to industrial pursuits and attendant patenting and licensing. Berman, "Why Did Universities Start Patenting?"; Gieryn, *Cultural Boundaries of Science*; Metlay, "Reconsidering Renormalization"; Shapin, *Scientific Life*.

23. Schacht, *Bayh-Dole Act*.

24. University technology transfer offices, which seek to license faculty and student research, influence whether and how synthetic biologists patent their work regardless of disciplinary background or personal stance toward ownership and sharing of research methods, tools, and products.

25. Oldham, Hall, and Burton, "Synthetic Biology."

26. Ibid.; Van Doren, Koenigstein, and Reiss, "Development of Synthetic Biology."

27. Rai and Boyle, "Synthetic Biology."

28. Thomas is a pseudonym.

29. By "ideas," Silver referenced an earlier topic of conversation, in which students asked about ways in which, for example, standards for measuring and manufacturing biological systems might be publishable.

30. In her ethnographic study of the copyist painters of Dafen, China, Winnie Wong points out that the aesthetic theories that underlie intellectual property are inaccurately universalist. She questions "on the one hand, the practices of the creative self required in producing creative work, and on the other, the erasure of authorship made possible through postmodernist appropriation and return." Such a claim reminds us that American copyright law depends on Euro-American assumptions about creative genius, novelty, and authenticity. Wong, *Van Gogh on Demand*, 32.

31. See chapter 1.

32. Hagstrom, *Scientific Community*.

33. Latour and Woolgar, *Laboratory Life*.

34. Frow, "Making Big Promises Come True?," 441.

35. Kelty, "Free Software / Free Science."

36. For a cultural history of copyleft licensing, see Kelty, *Two Bits*.

37. Mackenzie et al., "Classifying, Constructing, and Identifying Life Standards as Transformations of 'the Biological,'" 16.

38. Despite the BioBricks Foundation's best efforts, BioBricks would never become biology's ohm, or even its screw thread. Two problems quickly became apparent: first, not all synthetic biologists thought that it was possible to standardize living stuff; and second, those who did were not convinced that Knight's standard was the right one with which to do so. Beginning in 2003, various competing standards, each drafted by a different synthetic biology lab, began to proliferate—each new standard presented its own benefits and drawbacks, depending on the sort of engineering promulgated by the respective standard's proponents. Chris Anderson, Adam Arkin, and Jay Keasling, all at the University of California, Berkeley, proposed a new standard named "BglBricks" that would make it easier to perform automated protein fusions than do BioBricks. Other proposed standards included the "Biofusion" standard (implemented by Harvard's Silver Lab to avoid problematic frame shifts in DNA transcription), the University of Alberta's "BioBytes" assembly standard, the "Fusion Parts" standard developed by undergraduate students from Freiburg, Germany, and the "BioBricks++" standard (which hoped to improve on BioBricks by allowing for "scarless" assembly of multiple BioBrick parts).

39. The slogan "Free as in speech, not as in beer" clarifies the aims of FLOSS: while free speech is a right afforded citizens of liberal democracies, free beer is not.

40. A nonviral license, whether or not the inventor asserts rights, is by definition

"free." In part, this contract scheme is a reaction to the dominant proprietary landscape in biotechnology, which utilizes patents rather than copyright.

41. See https://biobricks.org/bpa/faq.

42. Hayden, *When Nature Goes Public*, 28–29 (emphasis in original).

43. Helmreich, "Species of Biocapital," 464. See also Helmreich, *Alien Ocean*, 106–44.

44. Interested readers can find a table summarizing JCVI's cipher here: http://www.righto.com/2010/06/using-arc-to-decode-venters-secret-dna.html.

45. "See things not as they are but as they might be" is a paraphrase of an admonition by Felix Adler, founder of the Ethical Culture Fieldston School, where J. Robert Oppenheimer was a student. This quotation was lifted from Bird and Sherwin, *American Prometheus*, 19.

46. "What I cannot build, I cannot understand." This sentence is a misquotation of Feynman's famous so-called "last blackboard" at Caltech, which reads, "What I cannot create, I do not understand." This quotation is much loved by synthetic biologists, who read it as an encapsulation of their overarching project, by which manufacturing living things expedites knowledge of life. See introduction.

47. Joyce, *Portrait of the Artist as a Young Man*, 200.

48. Carl Zimmer, "Copyright Law Meets Synthetic Life Meets James Joyce," *Discover* blog *The Loom*, March 15, 2011, http://blogs.discovermagazine.com/loom/2011/03/15/copyright-law-meets-synthetic-life-meets-james-joyce/#.V0NJv5N96fQ.

49. O'Connell, "Has James Joyce Been Set Free?" See also Ewalt, "Craig Venter's Genetic Typo."

50. Johns, *Nature of the Book*, 134.

51. Doyle, *On Beyond Living*; Kay, *Who Wrote the Book of Life?*; Keller, *Refiguring Life*; Keller, *Making Sense of Life*; Thacker, *Biomedia*.

52. Pottage, "Too Much Ownership," 152–53.

53. Johns, *Nature of the Book*, 517.

54. Ibid.

Interlude Four

1. Levin, " 'Tones from out of Nowhere,' " 34 (emphasis in original).

2. Adorno, "Music and New Music."

3. Borio, "New Technology, New Techniques."

4. Pinch and Trocco, *Analog Days*.

5. Ibid., 41.

6. Ibid., 281.

7. Bijsterveld and Pinch, " 'Should One Applaud?' "

8. Hütter and Schneider, interview with Kraftwerk.

9. Levin, " 'Tones from out of Nowhere,' " 62.

Chapter Four

1. Marx, *Karl Marx*, 368.
2. Witze, "Light in the Dark."
3. Noble, *Forces of Production*; Noble, *Progress without People*.
4. Bernard, *Lectures on the Phenomena of Life Common to Animals and Plants*, 259, quoted in Reynolds, "Cell's Journey," 67.
5. Reynolds, "Cell's Journey," 69.
6. As I have already shown, this is just one of many historical precedents that synthetic biologists call upon in describing their project (along with, e.g., synthetic chemistry, the rise of personal computing, and the Neolithic Revolution), even as they also declare their work to be radical and unprecedented (see chapter 2).
7. Hounshell, *From the American System to Mass Production*.
8. Berggren, "Lean Production"; Babson, *Lean Work*; Shimokawa, *Japanese Automobile Industry*.
9. "Go-Nowhere Generation"; "Is This Really the Worst Economic Recovery since the Depression?"; Tritch, "Still Crawling out of a Very Deep Hole"; D. Baker and Hassett, "Human Disaster of Unemployment"; Michael Cooper, "Many American Workers Are Underemployed and Underpaid"; "Economy Downshifts."
10. Caesar, Riese, and Seitz, "Betting on Biofuels"; *Codexis Announces Three-Year Extension of Collaboration Agreement with Merck*; *Codexis Enters into Research Collaboration with Schering-Plough*; *Codexis, Bristol-Myers Squibb Sign Research Agreement*; Rajagopal et al., "Recent Developments in Renewable Technologies"; "Shell's Brash Biofuels Partner"; Solecki, Scodel, and Epstein, "*Advanced Biofuel Market Report 2013*."
11. Sheridan, "Making Green," 1074.
12. Dyer-Witheford, *Cyber-Marx*, 221.
13. Dumit, "Prescription Maximization and the Accumulation of Surplus Health in the Pharmaceutical Industry."
14. Marx comments that when "the entire production process appears as not subsumed under the direct skillfulness of the worker, but rather as the technological application of science . . . [it is] the tendency of capital to give production a scientific character." The trend in synthetic biology I here identify doubly scientizes labor, as manual skill and tacit knowledge are gradually replaced by scientific instruments and the commodity being produced is the fruit of technoscientific labor. "Science too [is] 'among these productive forces'"; scientific knowledge and biotechnical products, furthermore, are the things to which such productive forces are applied. *Grundrisse*, 699.
15. Lazzarato, "From Capital-Labour to Capital-Life." Biagioli argues that "genius is copyright's *pharmakon*—simultaneously a cure and a poison." Biagioli, "Genius against Copyright," 1849. For the history of the attachment of "genius" to copyright law, see also Boyle, *Shamans, Software, and Spleens*; Coombe, *Cultural Life of Intellectual Properties*; Rose, *Authors and Owners*.

16. Lazzarato, "Immaterial Labor," 133.

17. Negri, "Constituent Republic," 89.

18. "Ginkgo Bioworks: Our Technology," http://ginkgobioworks.com/tech.html.

19. A brief quotation further encapsulates such thinking: one postdoc told me, "*E. coli* to me is more like engineering a Model T Ford, and yeast is like an aircraft carrier."

20. For example, a handout for an undergraduate synthetic biology lab course I attended in 2008 opened with the following explanation:

A car is a highly engineered system of interconnected parts. Many car parts are similar from car to car, but often the parts must be tailored to the size and function of the car. The chassis of a truck, a GTO muscle car and a Toyota hybrid are different, and so are many of the internal parts that make up the engine and the drive train. We might be able to move a radio from a truck chassis to a sports car chassis, but not much else. The car manufacturers are comfortable with this complexity, and it has little effect on the user of the car.

21. Similarly, Autodesk's Synthetic Biology Group promotes a four-part cycle. Carlos Olguin is a designer with a degree in information networking who now heads the Bio/Nano/Programmable Matter Group at Autodesk, the company best known for its CAD/CAM software. Olguin explained in a conference I attended in 2013, "Coming from a more mature engineering space, when we hear design, build, test, we scratch our heads and ask what about *simulating* before you build?"

22. Nye, *America's Assembly Line.*

23. Masami, "Myths of the Toyota System."

24. Schaffer, "Babbage's Intelligence," 222 (emphasis in original).

25. Mirowski, *Science-Mart*, 196.

26. For this reason, Bernadette Bensaude-Vincent has critiqued the International Genetically Engineered Machine (iGEM) competition for extracting value from undergraduate students: "Although the discourses about the iGEM competition are all about creativity, fun and excitement, the students provide a cheap way to fill the library of biobricks and to foster the process of cost reduction. Indeed students are not robots, but the replacement of technicians by automata has been very quick in DNA sequencing and the subsequent cost reduction over the past decade has been spectacular." Bensaude-Vincent, "Discipline-Building in Synthetic Biology," 128.

27. For a rich account of how the Toyota Way has inspired another contemporary discipline in the life sciences, consult H. Stevens, *Life out of Sequence.*

28. Brood, fleet, flock, culture, colony? To the best of my knowledge, there is no standardized English collective noun for a group of robots.

29. On "lab hands" and the craft of laboratory practice, see Doing, " 'Lab Hands' and the 'Scarlet O' "; Delamont and Atkinson, "Doctoring Uncertainty"; Mody, "Crafting the Tools of Knowledge"; Rasmussen, *Picture Control.*

30. Sheridan, "Making Green," 1074.

31. Ibid., 1075.

32. Hodgman and Jewett, "Cell-Free Synthetic Biology," 261.

33. Fortun, "Projecting Speed Genomics," 43.

34. See chapter 2.

35. Eunjung, "Scientists Now Creating 'App-Style' Life-Forms."

36. Haraway, *Simians, Cyborgs, and Women*, 166.

37. Dyer-Witheford, *Cyber-Marx*, 88.

38. Galison and Hevly, *Big Science*.

39. Venter has since transitioned into synthetic biology, and biofuels research in particular.

40. Preston, *Panic in Level 4*, 94.

41. Dyer-Witheford, *Cyber-Marx*, 223.

42. Eunjung, "Scientists Now Creating 'App-Style' Life-Forms."

43. This scientist spoke to me under condition of anonymity.

44. Notable examples of this literature include Melinda Cooper, *Life as Surplus*; Fortun, *Promising Genomics*; Franklin, *Dolly Mixtures*; Sunder Rajan, *Lively Capital*; Sunder Rajan, *Biocapital*; Waldby and Mitchell, *Tissue Economies*.

45. One answer comes from anthropologists of the global pharmaceutical market who argue that pharmaceutical companies extract surplus value from clinical trials, which are the machinery in which worker-participants "labor" by placing their bodies at risk for serious adverse drug reactions, even death. See, e.g., Dumit, *Drugs for Life*; Petryna, *When Experiments Travel*; Sunder Rajan, "Pharmaceutical Crises and Questions of Value." Another example is the physical and affective labor of Indian gestational surrogates exploited within globalized markets for assisted reproductive technologies. Vora, *Life Support*.

46. Stefan Helmreich critiques: "biological generativity is configured as accumulated labor power, the products of which can be harnessed to create productive futures. This belief is based, it bears emphasizing, on a metaphor: that organisms are laborers." Helmreich, "Species of Biocapital," 474.

47. Ibid. Landecker demonstrates how cells are commoditized within the material infrastructures of laboratories, from freezers to overnight shipping, as well as by the physical and bodily management of cellular animation and cessation that is necessary to coax cells into immortalized cell lines. Landecker, *Culturing Life*.

48. Oreskes and Conway, *Merchants of Doubt*; Proctor, *Cancer Wars*; Proctor and Schiebinger, *Agnotology*. Michael Betancourt defines agnotologic capitalism as "a capitalism systematically based on the production and maintenance of ignorance." Betancourt, "Immaterial Value and Scarcity in Digital Capitalism."

Interlude Five

1. Hooke, "Discourse of Earthquakes."

2. Hamilton, "ART. IX.-1 Artis Logicoe Rudimenta, with Illustrative Observations on Each Section," 236.

3. Daston, "Physicalist Tradition in Early Nineteenth Century French Geometry."

4. Kant, *Critique of Pure Reason*, 130.

5. Keller, "What Does Synthetic Biology Have to Do with Biology?," 292.

6. Kant, *Critique of Pure Reason*, 55.

Chapter Five

1. With apologies to Charis Thompson.

2. See http://www.nyu.edu/projects/xdesign/biotechhobbyist/.

3. See http://www.nyu.edu/projects/xdesign/biotechhobbyist/bio_about.html.

4. I own a similar shirt, a gift from a synthetic biologist. Cowell's depicts a mooing turtle; mine shows a dachshund with a speech bubble reading "cluck." Both T-shirts playfully demystify and defang the specter of transgenic organisms.

5. I here refer to them as "biohackers" rather than DIY biologists, because "hacker" avoids clear distinctions between professionals and laypeople, whereas "DIY biologist" confers on them an authority they neither possess nor seek.

6. Jackson, "Labour as Leisure," 61.

7. DIYbio LISTSERV, February 3, 2013.

8. The Personal Genome Project is described in more detail in the following chapter.

9. For a popular history of hackers and hacking, see Levy, *Hackers*.

10. Turner, *From Counterculture to Cyberculture*.

11. Synthetic biologists also compare their work to the HCC, as when Drew Endy remarked, "When Apple got started, there was [*sic*] already lots of commodity electronics. . . . Woz [Apple founder Steve Wozniak] didn't have to build his own power supply from stuff he dug up in the hills," thereby intimating that genetic engineering was, by comparison, proceeding piecemeal by "digging up" extant genetic material with which to cobble together recombinant organisms. Aldhous, "Redesigning Life."

12. Curry, "From Garden Biotech to Garage Biotech."

13. Moore, *Homebrew Computer Club Newsletter*.

14. A thermocycler is a laboratory apparatus that repeatedly raises and lowers the temperature in order to iteratively perform temperature-sensitive enzymatic reactions, most often amplifying (copying) DNA segments.

15. Drew Endy and Tom Knight have contributed to the DIY biology LISTSERV, as have their former students Jason Kelly, Reshma Shetty, and Austin Che, the founders of Ginkgo Bioworks. See chapter 4.

16. Levy, *Hackers*.

17. When I was working as a teaching assistant in the Department of Anthropology, one morning I asked the assembled undergraduates who should be applauded for the overnight hack that perched a fire truck atop MIT's dome. The students looked archly at one another, but no one admitted participation.

18. Vest, "MIT 141st Commencement Address," *MIT News*, http://news.mit.edu /2007/comm-vestspeech-0608.

19. See http://diybio.org/.

20. McCray, *Keep Watching the Skies!*

21. Phrenology, astrology, eugenics, orgone theory, mesmerism, creationism— Gordin reminds his readers that pseudoscience succeeds by mimicking scientific orthodoxy, and that "if you want to know what science is or has been, show me the contemporary pseudoscience." Pseudoscience is a useful object of study precisely because it promises to tell us what the status of science is in a particular political, historical, and cultural moment. Gordin, *Pseudoscience Wars*, 202, 3.

22. Popper, *Logic of Scientific Discovery*.

23. Wertheim, *Physics on the Fringe*.

24. As such, it is closer to an engineering field than to science, and it would be nonsensical to talk about "pseudoengineering" or "outsider engineering."

25. Beegan and Atkinson, "Professionalism, Amateurism and the Boundaries of Design," 308. For an example of how craft movements have called on antimodern-ist sensibilities to promote crafty self-sufficiency, see also Lears, *No Place of Grace*.

26. Cf. Gieryn, *Cultural Boundaries of Science*; Gieryn, "Boundary-Work and the Demarcation of Science from Non-science."

27. Science studies scholars demonstrate that nonprofessional research and labor have already been instrumental to biomedicine, in particular through the advocacy of patients and their families. See Epstein, *Impure Science*; Silverman, *Understanding Autism*.

28. For comparison, Nathan McCorkle, the most active member of the list, has posted over 2,600 messages. He recently earned a BS in biotechnology and has since founded a company that develops DNA synthesis devices. The eighth, ninth, and tenth most active members have posted significantly fewer messages, between 300 and 400 messages each over the last six years. Since it was established, I have archived and annotated over 30,000 emails sent among biohackers via this LISTSERV, data that supplement my ethnographic participant-observation.

29. Though not the myriad other oratorical and textual skills of self-presentation that successful scientists must learn.

30. Len Sassaman, e-mail to DIYbio LISTSERV, November 4, 2009.

31. It currently retails online for less than $50.

32. Codon Devices was the DNA synthesis company founded by George Church and Drew Endy, which shut its doors in April 2009 after failing to raise enough venture capital to stay afloat.

33. For an in-depth discussion of "the playful, tinkering nature" of DIYbio, see Calvert, "Engineering Biology and Society."

34. Mackenzie, "From Validating to Verifying."

35. Meyer, "Domesticating and Democratizing Science," 130.

36. Taubenberger et al., "Characterization of the 1918 Influenza Virus Polymerase Genes"; "1918 Flu Virus Is Resurrected."

37. Bennett et al., "From Synthetic Biology to Biohacking," 1111.

38. Kera, "Innovation Regimes Based on Collaborative and Global Tinkering," 29.

39. Mukunda, Oye, and Mohr, "What Rough Beast?"

40. Epstein, *Impure Science.*

41. Ellis and Waterton, "Environmental Citizenship in the Making."

42. Melamine is a toxic chemical that in 2007 and 2008 caused acute poisoning and death when sufficient concentrations adulterated animal feed and baby formula in mainland China.

43. Patterson, "Biopunk Manifesto."

44. E. Coleman, *Coding Freedom*, 4.

45. Kelty, *Two Bits.* Kelty's "recursive public" modifies Michael Warner's "public" to account for how people can organize themselves around not only shared discourses but also shared practices that enable their existence as a public. See also E. Coleman, "Political Agnosticism of Free and Open Source Software and the Inadvertent Politics of Contrast"; E. Coleman, "Hacker Politics and Publics"; E. Coleman, *Coding Freedom*; E. Coleman and Golub, "Hacker Practice."

46. All quotations are from the DIYbio LISTSERV.

47. With apologies to Max Weber.

48. R. Carlson and Brent, "DARPA Open-Source Biology." Endy's name does not appear on the letter, although Carlson reports that he helped draft it.

49. Ibid.

50. Ibid.

51. R. Carlson, "Open-Source Biology and Its Impact on Industry," 16.

52. S. Carlson, "Kitchen Counter DNA Lab."

53. Transhumanists are interested in technically augmenting or supplementing human bodies to maintain consciousness beyond the human life span, for example, by uploading consciousness online or via cryopreservation. Transhumanists maintain an active presence on the DIY biology LISTSERV.

54. At the next meeting, we would build gel boxes out of Legos, pour agar, and then use cut-up credit cards as combs around which wells formed as the gel hardened. We next hooked up the gel to a series of nine-volt batteries, running electricity through it in order to separate and analyze the previously isolated DNA.

55. Jacob, *Logic of Life.*

56. Kay, *Who Wrote the Book of Life?*

57. Oyama, *Ontogeny of Information.*

58. Schrödinger, *What Is Life?*

59. D. Adams, "Self Organization and Living Systems," quoted in Doyle, *On Beyond Living*, 35–38.

60. Helmreich, *Silicon Second Nature.*

61. I discuss this rhetoric in chapter 3.

62. On self-experimentation among cannabis biotechnologists, see Doyle, *Darwin's Pharmacy.* The subtitle of this section references his use of "first-person science."

63. Natasha Myers diagnoses a similar sort of excitability among protein crystallographers who recognize an "affective entanglement" lacing together the bodies of biologists and the biological substances with which they work. Myers, *Rendering Life Molecular*.

64. Novas and Rose, "Genetic Risk and the Birth of the Somatic Individual."

65. Rabinow, "Artificiality and Enlightenment."

66. Taussig, Rapp, and Heath, "Flexible Eugenics"; Rapp, Heath, and Taussig, "Genealogical Dis-ease."

67. Rapp, *Testing Women, Testing the Fetus*.

68. Smith, *Body of the Artisan*, 8.

69. For theorists of craft, see Becker, "Arts and Crafts"; Dormer, *Culture of Craft*; Greenhalgh, "History of Craft"; Metcalf, "Replacing the Myth of Modernism"; Paxson, *Life of Cheese*; Sennett, *Craftsman*. For accounts of the role of craft in scientific practice, consult, among others, H. Collins, "What Is Tacit Knowledge?"; Delamont and Atkinson, "Doctoring Uncertainty"; Myers, "Molecular Embodiments and the Body-Work of Modeling in Protein Crystallography"; Nutch, "Gadgets, Gizmos, and Instruments"; Polanyi, *Personal Knowledge*; Schaffer, "Glass Works"; Secord, "Science in the Pub"; Smith, *Body of the Artisan*.

70. The "Slow Science" movement is to knowledge production what "Slow Food" is to gustatory consumption. Digesting data and food both take time. The Slow Science movement's rallying cry is "Bear with us, while we think." To read their manifesto, visit the homepage of the Slow Science Academy: http://slow-science.org/.

Interlude Six

1. Agassiz, *Twelve Lectures on Comparative Embryology*, 432; Agassiz, *Essay on Classification*, 178.

2. Keller, "What Does Synthetic Biology Have to Do with Biology?," 299 (emphasis in original).

3. Keller, *Making Sense of Life*, 22.

4. Ibid., 8.

5. Ray, "Evolutionary Approach to Synthetic Biology," 195. See also Helmreich, *Silicon Second Nature*.

6. Sinsheimer, "Recombinant DNA—on Our Own."

7. Roblin, "Synthetic Biology," 172.

Chapter Six

1. Pierce, "Shirley Temple Three," 89.

2. In a companion interview about his short story, Pierce traced a long history

of American fascination with woolly mammoths, reporting that "Thomas Jefferson hoped [mammoths] weren't actually extinct and told Lewis and Clark to be on the lookout for a live one on their way to the west coast." Leyshon, "This Week in Fiction."

3. Ibid.

4. Here it is useful to differentiate between clones, chimeras, and hybrids. Technically speaking, *clones* are genetic replicas of one another. While Dolly the Sheep (born at the Roslin Institute in 1996) and other products of somatic cell nuclear transfer (SCNT, a procedure I describe in more detail below) are popularly described as "clones," this terminology is inaccurate, strictly speaking. Animals resulting from SCNT are genetic *chimeras*. Bearing nuclear DNA from one animal and mitochondrial DNA (mtDNA) from a donated egg, they are sometimes gestated by a third animal. A *hybrid*, in contrast, is an animal resulting from sexual reproduction of two organisms from different species (e.g., a mule).

5. Friese, "Models of Cloning, Models for the Zoo," 369.

6. Ibid., 379n20.

7. The quagga is an extinct relative of the zebra, once found in the southern regions of South Africa but wiped out by overhunting.

8. Turner, *From Counterculture to Cyberculture.*

9. Wakayama et al., "Production of Healthy Cloned Mice from Bodies Frozen at −20°C for 16 Years," 17318.

10. Church and Regis, *Regenesis*, 9.

11. Franklin, *Dolly Mixtures*, 41.

12. Zimmer, "Bringing Them Back to Life," 30 (emphasis added).

13. Ibid.

14. Benson, "Endangered Birds and Epistemic Concerns."

15. Haraway, *Primate Visions*, 45.

16. Conservation biologists like Archer are aware that reviving a single individual of an extinct species is not enough: to ensure its survival a second time around, they must engineer social groups of these animals and study their behavior and habitats. To learn more about thylacine behavior, Archer worked with a local nonagenarian named Peter Carter, whose father and brother had hunted thylacines in the 1920s. Carter had kept two of them as pets in the small cabin he shared with his family. As he and Archer walked around the land that had once been his family's, Carter cried as he remembered waking in the night as a young boy and hearing the yip and wail of thylacines circling the cabin.

17. I did not.

18. Parry, "Technologies of Immortality."

19. Landecker, *Culturing Life*, 227–28.

20. Ibid., 154.

21. Malinowski, *Argonauts of the Western Pacific*, xv.

22. Radin, "Latent Life," 487.

23. The term "latent life" is often attributed to experimental physiologist Claude Bernard in his 1878 *Lectures on the Phenomena of Life Common to Animals and Plants* to denote "chemical indifference" induced by low temperatures. See Perlman, "Concept of the Organism in Physiology." Eight years earlier, in his *Hereditary Genius*, eugenicist Francis Galton referenced "latent life" without defining the term. "Latent life" was simultaneously revived in 1910 and 1911 by biologists Stéphane Leduc and Alexis Carrel. Leduc describes "latent life" as akin to "a machine that has been stopped, but which retains its form and substance unaltered, and may be started again whenever the obstacle to its progress is removed." Leduc, *Mechanism of Life*, 7. In reporting on grafts of arterial tissue, Carrel explains, "a tissue is in latent life when its metabolism becomes so slight that it cannot be detected, and also when its metabolism is completely suspended. Latent life means, therefore, two different conditions, unmanifested actual life and potential life." Carrel, "Latent Life of Arteries," 460. The phrase was again resurrected by entomologist David Keilin in 1958 to denote resuscitating organisms kept in suspension by desiccation or low temperatures. Keilin, "Leeuwenhoek Lecture." For a history of cryptobiosis, see Roosth, "Life, Not Itself."

24. Reardon, *Race to the Finish*.

25. Walker and Stiles, "Consequences of Legal Ivory Trade," author reply, 1634–35; Stokstad, "Conservation Biology"; Wasser et al., "Conservation."

26. Chakrabarty, *Provincializing Europe*, 243.

27. Dinshaw, *How Soon Is Now?*, 4–5 (emphasis in original).

28. Church and Regis, *Regenesis*, 12.

29. Carr and Church, "Genome Engineering."

30. Wang et al., "Programming Cells by Multiplex Genome Engineering and Accelerated Evolution."

31. Church and Regis, *Regenesis*, 148.

32. Kelty, *Two Bits*, 89.

33. Fortun, "Human Genome Project and the Acceleration of Biotechnology"; Fortun, *Promising Genomics*; Sunder Rajan, *Biocapital*.

34. I address literature on biocapital in more detail in chapters 3 and 4. Michael Fortun writes of the shared valences of scientific "hype" (exaggerated publicizing and future-oriented promising) and "hyperbole": "what hype is must be speculated on, an operation that thus partakes of the rhetoric of hyperbole itself." Fortun, *Promising Genomics*, 304n1.

35. Kaushik Sunder Rajan, following Max Weber, describes the biotech start-up phenomenon as symptomatic of salvationary messianic thinking: "to generate value in the present [is] to make a certain kind of future possible. . . . Excess, expenditure, exuberance, risk, and gambling can be generative because they can create that which is unanticipated, perhaps even unimagined." Sunder Rajan, *Biocapital*, 116. Such abductive reasoning, Stefan Helmreich claims, "joins hope to reason, present texts to future contexts, contemporary life forms to scientific forms of life yet to come." Helmreich, *Alien Ocean*, 172. For further scholarship on "biocapital," see

also Franklin and Lock, *Remaking Life and Death*; Helmreich, "Species of Biocapital"; Thompson, "Biotech Mode of Reproduction."

36. Science studies scholars have diagnosed hope and anticipation as technoscientific affective states colored by both desire and anxiety: "Anticipation reconfigures the 'lay of the land' as sites that in colonial logics were mapped as either primitive (past and out of time) or modern (present and in time) and turns them both into productive ground for anticipatory interventions." V. Adams, Murphy, and Clarke, "Anticipation," 248.

37. Thompson, *Making Parents*, 271–74.

38. Ritvo, "Beasts in the Jungle (or Wherever)."

39. Zimmer, "Bringing Them Back to Life," 33.

40. Ibid., 41.

41. Ritvo, "Race, Breed, and Myths of Origin," 11, 14.

42. P. Stevens, "Species," 305.

43. Mayr, *Systematics and the Origin of Species, from the Viewpoint of a Zoologist*.

44. Friese, "Models of Cloning, Models for the Zoo," 384.

45. Cussins, "Confessions of a Bioterrorist," 191.

46. Friese, "Classification Conundrums," 147.

47. In *The Platypus and the Mermaid*, Harriet Ritvo describes how Victorian taxonomic practices reflected the myriad relations between humans and animals—as sources of food, amusement, and labor and as proxies for human characteristics. "The classification of animals, like that of any group of significant objects, is apt to tell as much about the classifiers as about the classified" (xii).

48. About 20 percent of the DNA extracted was identified as archaean, viral, or eukaryotic; the rest was unidentifiable. James Watson's genome was the baseline human DNA sequence against which the mammoth DNA was compared as a control.

49. Zimov, "Pleistocene Park," 796.

50. Fabian, *Time and the Other*.

51. Church and Regis, *Regenesis*, 149.

52. Cronon, "The Trouble with Wilderness," 16.

53. Landecker, *Culturing Life*.

54. Marx, "18th Brumaire of Louis Bonaparte," quoted in Chakrabarty, *Provincializing Europe*, 245.

55. Boym, *Future of Nostalgia*, xv.

Conclusion

1. Salingaros and Alexander, *Anti-architecture and Deconstruction*.

2. See http://openwetware.org/wiki/Endy:Research.

3. Daston and Galison, *Objectivity*, 46.

4. For a more sustained argument that such efforts at times embed Platonic thinking, see Roosth, "Life, Not Itself."

5. Riskin, *Genesis Redux*, 14.

6. Helmreich, *Silicon Second Nature*.

Appendix

1. Malinowski, *Argonauts of the Western Pacific*.

2. Snow, *Two Cultures*.

3. A nucleoside is a nucleotide lacking a phosphate group.

4. Unless otherwise noted, I use scientists' real names rather than pseudonyms. All professors are referred to by their real names, as it would be impossible to anonymize those researchers who have been profiled by the popular press. I use the first or full name of all graduate students, undergraduates, and other researchers who granted me permission to do so. If a student preferred to be anonymized or in the rare case in which I could not reach that person, I have employed a pseudonym.

5. Indeed, as I complete this book, synthetic biology has begun to infect pop culture, and not just via the usual channels of pontifical TED talks, fawning *New Yorker* articles, and polished NPR interviews. An installation of art inspired by synthetic biology was exhibited in 2013, to widespread adulation, at the Science Gallery at Trinity College Dublin. The algorithm of my Spotify account recommended last week that I listen to a song titled "Venter." It is, I was surprised to discover, in fact inspired by the J. Craig Venter Institute's synthetic *Mycoplasma*. And while recently watching season 2 of the Canadian SF television series *Orphan Black,* I was rudely snapped out of my suspension of disbelief when evolutionary biologist and genetic clone Cosima explained to her geneticist girlfriend that her genome contained a genetic "watermark" similar to those the J. Craig Venter Institute encoded in their synthetic genomes.

6. "What's in a Name?," 1073.

Bibliography

Adams, D. H. "Self Organization and Living Systems: Is DNA Artificial Intelligence?" *Medical Hypotheses* 29 (1989): 223–29.

Adams, Vincanne, Michelle Murphy, and Adele Clarke. "Anticipation: Technoscience, Life, Affect, Temporality." *Subjectivity* 28 (2009): 246–65.

Adorno, Theodor W. *Minima Moralia: Reflections on a Damaged Life.* Translated by E. F. N. Jephcott. London: Verso, 2005.

———. "Music and New Music." In *Quasi Una Fantasia: Essays on Modern Music,* 249–68. London: Verso, 1998.

Agassiz, Louis. *An Essay on Classification.* London: Longman, Brown, Green, Longmans, and Roberts, 1859.

———. *Twelve Lectures on Comparative Embryology: Delivered before the Lowell Institute, in Boston, December and January, 1848–9.* Boston: Henry Flanders, 1849.

Aldhous, Peter. "Redesigning Life: Meet the Bio-Hackers." *New Scientist,* May 20, 2006, 43–47.

Altman, Lawrence K. "Michael DeBakey, Rebuilder of Hearts, Dies at 99." *New York Times,* July 13, 2008, sec. Health.

Amato, Ivan. *Stuff: The Materials the World Is Made Of.* New York: Basic Books, 1997.

Anderson, Chris. "The End of Theory: The Data Deluge Makes the Scientific Method Obsolete." *Wired,* June 2008, 108–9.

Andrianantoandro, Ernesto, Subhayu Basu, David K. Karig, and Ron Weiss. "Synthetic Biology: New Engineering Rules for an Emerging Discipline." *Molecular Systems Biology* 2 (2006): 1–14. http://www.ncbi.nlm.nih.gov/pmc/articles/PMC1681505/.

Ansell-Pearson, Keith. *Viroid Life: Perspectives on Nietzsche and the Transhuman Condition.* New York: Routledge, 1997.

Ashman, Keith. *After the Science Wars: Science and the Study of Science.* New York: Routledge, 2001.

Atwood, Margaret. *Oryx and Crake: A Novel*. New York: Random House, 2004.

Babson, Steve. *Lean Work: Empowerment and Exploitation in the Global Auto Industry*. Detroit: Wayne State University Press, 1995.

Bahnisch, Mark. "Embodied Work, Divided Labour: Subjectivity and the Scientific Management of the Body in Frederick W. Taylor's 1907 'Lecture on Management.'" *Body and Society* 6, no. 1 (2000): 51–68.

Baker, Dean, and Kevin Hassett. "The Human Disaster of Unemployment." *New York Times*, May 12, 2012, sec. Opinion / Sunday Review.

Baker, Monya. "Big Biology: The 'omes Puzzle." *Nature* 494, no. 7438 (February 27, 2013): 416–19.

Bakshi-Doron, Eliyahu. *Binyan Av*. Vol. 4, *siman* 43. Jerusalem: Mekhon "Binyan Av," 2002.

Baldwin, Geoff. *Synthetic Biology: A Primer*. London: Imperial College Press, 2012.

Balmer, Andrew S., and Kate J. Bulpin. "Left to Their Own Devices: Post-ELSI, Ethical Equipment and the International Genetically Engineered Machine (iGEM) Competition." *BioSocieties* 8, no. 3 (2013): 311–35.

Bamford, Sandra C. *Biology Unmoored: Melanesian Reflections on Life and Biotechnology*. Berkeley: University of California Press, 2007.

Bardini, Thierry. *Junkware*. Minneapolis: University of Minnesota Press, 2011.

Barr, Alfred Hamilton. *Cubism and Abstract Art: Painting, Sculpture, Constructions, Photography, Architecture, Industrial Art, Theatre, Films, Posters, Typography*. Cambridge, MA: Harvard University Press, 1936.

Becker, Howard S. "Arts and Crafts." *American Journal of Sociology* 83, no. 4 (1978): 862–89.

Beegan, Gerry, and Paul Atkinson. "Professionalism, Amateurism and the Boundaries of Design." *Journal of Design History* 21, no. 4 (2008): 305–13.

Belasco, Warren. "Algae Burgers for a Hungry World? The Rise and Fall of Chlorella Cuisine." *Technology and Culture* 38, no. 3 (1997): 608–34.

———. "Future Notes: The Meal-in-a-Pill." *Food and Foodways* 8, no. 4 (2000): 253–71.

———. *Meals to Come: A History of the Future of Food*. Berkeley: University of California Press, 2006.

Benner, Steven A., and A. Michael Sismour. "Synthetic Biology." *Nature Reviews: Genetics* 6, no. 7 (2005): 533–43.

Bennett, Gaymon, Nils Gilman, Anthony Stavrianakis, and Paul Rabinow. "From Synthetic Biology to Biohacking: Are We Prepared?" *Nature Biotechnology* 27, no. 12 (December 2009): 1109–11.

Bensaude-Vincent, Bernadette. "The Construction of a Discipline: Materials Science in the United States." *Historical Studies in the Physical and Biological Sciences* 31, no. 2 (2001): 223–48.

———. "Discipline-Building in Synthetic Biology." *Studies in History and Philosophy of Biological and Biomedical Sciences* 44, no. 2 (2013): 122–29.

———. "Synthetic Biology as a Replica of Synthetic Chemistry? Uses and Misuses of History." *Biological Theory* 4, no. 4 (2009): 314–18.

Bensaude-Vincent, Bernadette, and Isabelle Stengers. *A History of Chemistry.* Translated by Deborah Van Dam. Cambridge, MA: Harvard University Press, 1996.

Benson, Etienne. "Endangered Birds and Epistemic Concerns: The California Condor." In *Endangerment, Biodiversity, and Culture,* edited by Fernando Vidal and Nélia Días, 174–94. New York: Routledge, 2015.

Berggren, C. "Lean Production—the End of History?" *Work, Employment, and Society* 7, no. 2 (1993): 163–88.

Berman, Elizabeth Popp. "Why Did Universities Start Patenting? Institution-Building and the Road to the Bayh-Dole Act." *Social Studies of Science* 38, no. 6 (2008): 835–71.

Bernal, J. D. "Definitions of Life." *New Scientist* 23 (1967): 12–14.

Bernard, Claude. *Lectures on the Phenomena of Life Common to Animals and Plants.* Vol. 1. Translated by Hebbel E. Hoff, Roger Guillemin, and Lucienne Guillemin. Springfield, IL: Charles C. Thomas, 1974. First published 1878.

Bernstein, Mary, and Renate Reimann, eds. *Queer Families, Queer Politics.* New York: Columbia University Press, 2001.

Betancourt, Michael. "Immaterial Value and Scarcity in Digital Capitalism." *CTheory,* June 10, 2010. http://www.ctheory.net/articles.aspx?id=652#_edn14.

Biagioli, Mario. "Between Knowledge and Technology: Patenting Methods, Rethinking Materiality." *Anthropological Forum* 22, no. 3 (2012): 285–99.

———. "Genius against Copyright: Revisiting Fichte's Proof of the Illegality of Reprinting." *Notre Dame Law Review* 86, no. 5 (2011): 1847–68.

Bijsterveld, Karin, and Trevor J. Pinch. " 'Should One Applaud?' Breaches and Boundaries in the Reception of New Technology in Music." *Technology and Culture* 44, no. 3 (2003): 536–59.

Bird, Kai, and Martin J. Sherwin. *American Prometheus: The Triumph and Tragedy of J. Robert Oppenheimer.* New York: Random House, 2005.

Blaszczyk, Regina Lee. "Styling Synthetics: DuPont's Marketing of Fabrics and Fashions in Postwar America." *Business History Review* 80, no. 3 (2006): 485–528.

Borio, Gianmario. "New Technology, New Techniques: The Aesthetics of Electronic Music in the 1950's." *Interface* 22, no. 1 (1993): 77–87.

Bowler, Peter J. "Darwinism and the Argument from Design: Suggestions for a Reevaluation." *Journal of the History of Biology* 10, no. 1 (1977): 29–43.

———. *The Non-Darwinian Revolution: Reinterpreting a Historical Myth.* Baltimore: Johns Hopkins University Press, 1992.

Boyle, James. *Shamans, Software, and Spleens: Law and the Construction of the Information Society.* Cambridge, MA: Harvard University Press, 1996.

Boym, Svetlana. *The Future of Nostalgia.* New York: Basic Books, 2001.

Brain, Robert. *The Pulse of Modernism: Physiological Aesthetics in Fin-de-Siècle Europe.* Seattle: University of Washington Press, 2015.

Brandt, Allan. *The Cigarette Century: The Rise, Fall, and Deadly Persistence of the Product That Defined America.* New York: Basic Books, 2009.

Browne, Janet. *Charles Darwin: A Biography.* Vol. 1, *Voyaging.* Princeton, NJ: Princeton University Press, 1996.

———. *Charles Darwin: A Biography.* Vol. 2, *The Power of Place.* Princeton, NJ: Princeton University Press, 2003.

Bugos, Glenn E., and Daniel J. Kevles. "Plants as Intellectual Property: American Practice, Law, and Policy in World Context." *Osiris* 7 (1992): 74–104.

Burroughs, William. *The Soft Machine.* New York: Grove/Atlantic, 2011.

Butler, Judith. *Bodies That Matter: On the Discursive Limits of Sex.* New York: Routledge, 2011. First published 1993.

Caesar, William K., Jens Riese, and Thomas Seitz. "Betting on Biofuels." *McKinsey Quarterly,* no. 2 (2007): 53–63.

Calvert, Jane. "The Commodification of Emergence: Systems Biology, Synthetic Biology and Intellectual Property." *BioSocieties* 3, no. 4 (2008): 383–98.

———. "Engineering Biology and Society: Reflections on Synthetic Biology." *Science, Technology, and Society* 18, no. 3 (2013): 405–20.

———. "Synthetic Biology: Constructing Nature?" *Sociological Review* 58 (2010): 95–112.

Carlson, Robert. "Open-Source Biology and Its Impact on Industry." *IEEE Spectrum* 38, no. 5 (2001): 15–17.

———. "The Pace and Proliferation of Biological Technologies." *Biosecurity and Bioterrorism: Biodefense Strategy, Practice, and Science* 1, no. 3 (2003): 203–14.

Carlson, Robert, and Roger Brent. "DARPA Open-Source Biology." October 2000. synthesis.cc/DARPA_OSB_Letter.pdf.

Carlson, Shawn. "Kitchen Counter DNA Lab." *Make Magazine,* January 2007.

Carr, Peter A., and George M. Church. "Genome Engineering." *Nature Biotechnology* 27, no. 12 (2009): 1151–62.

Carrel, Alexis. "Latent Life of Arteries," *Journal of Experimental Medicine* 12, no. 4 (1910): 460–86.

Carroll, Joe, Rachel Donnelly, and Kevin O'Sullivan. "'We Are Learning Language in Which God Created Life.'" *Irish Times,* June 27, 2000.

Carson, Rachel. *Silent Spring.* Boston: Houghton Mifflin, 2002. First published 1962.

Cartwright, Lisa. "'Experiments of Destruction': Cinematic Inscriptions of Physiology." *Representations,* no. 40 (1992): 129–52.

Chadarevian, Soraya de. *Designs for Life: Molecular Biology after World War II.* Cambridge: Cambridge University Press, 2002.

———. "Graphical Method and Discipline: Self-Recording Instruments in Nineteenth-Century Physiology." *Studies in History and Philosophy of Science Part A* 24, no. 2 (1993): 267–91.

Chadarevian, Soraya de, and Nick Hopwood, eds. *Models: The Third Dimension of Science.* Stanford, CA: Stanford University Press, 2004.

Chakrabarty, Dipesh. *Provincializing Europe: Postcolonial Thought and Historical Difference*. Princeton, NJ: Princeton University Press, 2007.

Chan, Leon Y., Sriram Kosuri, and Drew Endy. "Refactoring Bacteriophage T7." *Molecular Systems Biology* 1, no. 1 (2005): E1–10.

Chatton, E. P. L. *Titres et travaux scientifiques (1906–1937) de Edouard Chatton*. Sottano, Italy: Sette, 1938.

Church, George. "Constructive Biology." Paper presented at "The Edge: The Third Culture," Cambridge, MA, June 26, 2006. http://www.edge.org/3rd_culture/church06/church06_index.html.

———. "Let Us Go Forth and Safely Multiply." *Nature* 438, no. 7067 (November 24, 2005): 423.

Church, George M., and Ed Regis. *Regenesis: How Synthetic Biology Will Reinvent Nature and Ourselves*. New York: Basic Books, 2012.

Clifford, James, and George E. Marcus, eds. *Writing Culture: The Poetics and Politics of Ethnography*. Berkeley: University of California Press, 1986.

Clinton, Bill. "Reading the Book of Life: Text of the White House Statements on the Human Genome Project." *New York Times*, June 27, 2000, sec. Science.

"Codexis Announces Broad Technology Collaboration with Pfizer." Press release. Redwood City, CA, July 26, 2004. http://ir.codexis.com/phoenix.zhtml?c=208899&p=irol-newsArticle_pf&ID=1181037.

"Codexis Announces Three-Year Extension of Collaboration Agreement with Merck." Press release. Redwood City, CA, May 22, 2012. http://ir.codexis.com/phoenix.zhtml?c=208899&p=irol-newsArticle_pf&ID=1698577.

"Codexis, Bristol-Myers Squibb Sign Research Agreement." Press release. Redwood City, CA, May 24, 2005. http://ir.codexis.com/phoenix.zhtml?c=208899&p=irol-newsArticle&ID=1181053.

"Codexis Enters into Research Collaboration with Schering-Plough." Press release. Redwood City, CA, March 21, 2006. http://ir.codexis.com/phoenix.zhtml?c=208899&p=irol-newsArticle_pf&ID=1181063.

Coleman, E. Gabriella. *Coding Freedom: The Ethics and Aesthetics of Hacking*. Princeton, NJ: Princeton University Press, 2013.

———. "Hacker Politics and Publics." *Public Culture* 23, no. 3 (2011): 511–16.

———. "The Political Agnosticism of Free and Open Source Software and the Inadvertent Politics of Contrast." *Anthropological Quarterly* 77, no. 3 (2004): 507–19.

Coleman, E. Gabriella, and Alex Golub. "Hacker Practice: Moral Genres and the Cultural Articulation of Liberalism." *Anthropological Theory* 8, no. 3 (2008): 255–77.

Coleman, William. *Biology in the Nineteenth Century: Problems of Form, Function, and Transformation*. Cambridge: Cambridge University Press, 1977.

Coleman, William, and Frederic Lawrence Holmes. *The Investigative Enterprise: Experimental Physiology in Nineteenth-Century Medicine*. Berkeley: University of California Press, 1988.

Collins, Francis. "Has the Revolution Arrived?" *Nature* 464, no. 7289 (2010): 674–75.

Collins, Harry M. "What Is Tacit Knowledge?" In *The Practice Turn in Contemporary Theory*, edited by Theodore R. Schatzki, Karin Knorr-Cetina, and Eike von Savigny, 107–19. New York: Routledge, 2001.

Coombe, Rosemary J. *The Cultural Life of Intellectual Properties: Authorship, Appropriation, and the Law*. Durham, NC: Duke University Press, 1998.

Cooper, Melinda. *Life as Surplus: Biotechnology and Capitalism in the Neoliberal Era*. Seattle: University of Washington Press, 2008.

Cooper, Michael. "Many American Workers Are Underemployed and Underpaid." *New York Times*, June 18, 2012, sec. U.S.

Copeland, Herbert Faulkner. *The Classification of Lower Organisms*. Palo Alto, CA: Pacific Books, 1956.

Cottington, David. *Cubism and Its Histories*. Manchester, UK: Manchester University Press, 2004.

Creager, Angela N. H. *The Life of a Virus: Tobacco Mosaic Virus as an Experimental Model, 1930–1965*. Chicago: University of Chicago Press, 2002.

Cronon, William. "The Trouble with Wilderness; or, Getting Back to the Wrong Nature." *Environmental History* 1, no. 1 (1996): 7–28.

Cunningham, Andrew, and Nicholas Jardine. *Romanticism and the Sciences*. Cambridge: Cambridge University Press, 1990.

Curry, Helen. "From Garden Biotech to Garage Biotech: Amateur Experimental Biology in Historical Perspective." *British Journal for the History of Science* 47, no. 3 (2014): 539–65.

Cussins, Charis Thompson. "Confessions of a Bioterrorist: Subject Position and Reproductive Technologies." In *Playing Dolly: Technocultural Formations, Fantasies, and Fictions of Assisted Reproduction*, edited by Elizabeth Ann Kaplan and Susan Merrill Squier, 189–219. New Brunswick, NJ: Rutgers University Press, 1999.

Daston, Lorraine. "The Physicalist Tradition in Early Nineteenth Century French Geometry." *Studies in History and Philosophy of Science Part A* 17, no. 3 (1986): 269–95.

Daston, Lorraine, and Peter Galison. *Objectivity*. New York: Zone Books, 2007.

Daston, Lorraine, and Katharine Park. *Wonders and the Order of Nature, 1150–1750*. New York: Zone Books, 1998.

Davis, Bernard D. "The Human Genome and Other Initiatives." *Science* 249, no. 4967 (1990): 342–43.

Dear, Peter. "What Is the History of Science the History Of? Early Modern Roots of the Ideology of Modern Science." *Isis* 96, no. 3 (2005): 390–406.

Delamont, Sara, and Paul Atkinson. "Doctoring Uncertainty: Mastering Craft Knowledge." *Social Studies of Science* 31, no. 1 (2001): 87–107.

Delebecque, Camille. "Bacteria, Archaea, Eukarya . . . + Synthetica?" http://blog .camilledelebecque.com/post/983925371/bacteria-archaea-eukarya-synthetica.

DeLillo, Don. *White Noise*. New York: Penguin, 1999.

Dinshaw, Carolyn. *How Soon Is Now? Medieval Texts, Amateur Readers, and the Queerness of Time*. Durham, NC: Duke University Press, 2012.

Doing, Park. "'Lab Hands' and the 'Scarlet O': Epistemic Politics and (Scientific) Labor." *Social Studies of Science* 34, no. 3 (2004): 299–323.

Dormer, Peter, ed. *The Culture of Craft*. Manchester, UK: Manchester University Press, 1997.

Douglas, Mary. *Purity and Danger: An Analysis of Concepts of Pollution and Taboo*. New York: Routledge, 1966.

Doyle, Richard. *Darwin's Pharmacy: Sex, Plants, and the Evolution of the Noösphere*. Seattle: University of Washington Press, 2011.

———. *On Beyond Living: Rhetorical Transformations of the Life Sciences*. Stanford, CA: Stanford University Press, 1997.

Dumit, Joseph. *Drugs for Life: How Pharmaceutical Companies Define Our Health*. Durham, NC: Duke University Press, 2012.

———. "Prescription Maximization and the Accumulation of Surplus Health in the Pharmaceutical Industry: The_Biomarx_Experiment." In *Lively Capital: Biotechnologies, Ethics, and Governance in Global Markets*, edited by Kaushik Sunder Rajan, 45–92. Durham, NC: Duke University Press, 2012.

Dunn, John J., F. William Studier, and M. Gottesman. "Complete Nucleotide Sequence of Bacteriophage T7 DNA and the Locations of T7 Genetic Elements." *Journal of Molecular Biology* 166, no. 4 (1983): 477–535.

Dyer-Witheford, Nick. *Cyber-Marx: Cycles and Circuits of Struggle in High-Technology Capitalism*. Urbana: University of Illinois Press, 1999.

"The Economy Downshifts." *New York Times*, April 29, 2012, sec. Opinion.

Edelman, Lee. *No Future: Queer Theory and the Death Drive*. Durham, NC: Duke University Press, 2004.

Ellis, Rebecca, and Claire Waterton. "Environmental Citizenship in the Making: The Participation of Volunteer Naturalists in UK Biological Recording and Biodiversity Policy." *Science and Public Policy* 31 (2004): 95–105.

Endersby, Jim. *Imperial Nature: Joseph Hooker and the Practices of Victorian Science*. Chicago: University of Chicago Press, 2008.

Endy, Andrew "Drew" David. "Development and Application of a Genetically-Structured Simulation for Bacteriophage T7." PhD diss., Dartmouth College, 1997.

Endy, Drew, Bonnie Bassler, and Rob Carlson. *Overview and Context of the Science and Technology of Synthetic Biology*. Presidential Commission for the Study of Bioethical Issues, Washington, DC, July 8, 2010. http://bioethics.gov/node/164.

Endy, Drew, and Roger Brent. "Modelling Cellular Behaviour." *Nature* 409, no. 6818 (2001): 391–95.

Epstein, Steven. *Impure Science: AIDS, Activism, and the Politics of Knowledge*. Berkeley: University of California Press, 1998.

Eunjung, Ariana. "Scientists Now Creating 'App-Style' Life-Forms." *Japan Times Online*, October 27, 2013. http://www.japantimes.co.jp/news/2013/10/27/world /scientists-now-creating-app-style-life-forms/.

Ewalt, David M. "Craig Venter's Genetic Typo." *Forbes*, March 14, 2011. http:// www.forbes.com/sites/davidewalt/2011/03/14/craig-venters-genetic-typo/.

Fabian, Johannes. *Time and the Other: How Anthropology Makes Its Object*. New York: Columbia University Press, 1983.

Farber, Paul Lawrence. *Finding Order in Nature: The Naturalist Tradition from Linnaeus to E. O. Wilson*. Baltimore: Johns Hopkins University Press, 2000.

Feferman, Linda. *Simple Rules . . . Complex Behavior*. Santa Fe, NM, 1992. Video.

Feynman, Richard. "Feynman's Office; the Last Blackboards." *Physics Today* 42, no. 2 (1989): 88.

Findlen, Paula. *Athanasius Kircher: The Last Man Who Knew Everything*. New York: Routledge, 2004.

———. *Possessing Nature: Museums, Collecting, and Scientific Culture in Early Modern Italy*. Berkeley: University of California Press, 1994.

Finlay, Susanna. "Engineering Biology? Exploring Rhetoric, Practice, Constraints and Collaborations within a Synthetic Biology Research Centre." *Engineering Studies* 5, no. 1 (2013): 26–41.

Forster, Anthony C., and George M. Church. "Synthetic Biology Projects in Vitro." *Genome Research* 17, no. 1 (2007): 1–6.

Fortun, Michael. "The Human Genome Project and the Acceleration of Biotechnology." In *Private Science: Biotechnology and the Rise of the Molecular Sciences*, edited by Arnold Thackray, 182–201. Philadelphia: University of Pennsylvania Press, 1998.

———. "Projecting Speed Genomics." In *The Practices of Human Genetics*, edited by Michael Fortun and Everett Mendelsohn, 25–48. London: Kluwer, 1999.

———. *Promising Genomics: Iceland and deCODE Genetics in a World of Speculation*. Berkeley: University of California Press, 2008.

Foucault, Michel. *The Order of Things: An Archaeology of the Human Sciences*. New York: Pantheon Books, 1971.

Francoeur, Eric. "The Forgotten Tool: The Design and Use of Molecular Models." *Social Studies of Science* 27, no. 1 (1997): 7–40.

Franklin, Sarah. "Biologization Revisited: Kinship Theory in the Context of the New Biologies." In *Relative Values: Reconfiguring Kinship Studies*, edited by Sarah Franklin and Susan McKinnon, 302–28. Durham, NC: Duke University Press, 2001.

———. *Dolly Mixtures: The Remaking of Genealogy*. Durham, NC: Duke University Press, 2007.

Franklin, Sarah, and Margaret Lock, eds. *Remaking Life and Death: Toward an Anthropology of the Biosciences*. Santa Fe, NM: School of American Research Press, 2003.

Friedmann, Herbert Claus. "From Butyribacterium to *E. coli*: An Essay on Unity in Biochemistry." *Perspectives in Biology and Medicine* 47, no. 1 (2004): 47–66.

Friese, Carrie. "Classification Conundrums: Categorizing Chimeras and Enacting Species Preservation." *Theory and Society* 39, no. 2 (2010): 145–72.

———. "Models of Cloning, Models for the Zoo: Rethinking the Sociological Significance of Cloned Animals." *BioSocieties* 4, no. 4 (2009): 367–90.

Frohlich, Xaq. "Accounting for Taste : Regulating Food Labeling in the 'Affluent Society,' 1945–1995." PhD diss., Massachusetts Institute of Technology, 2011.

Frow, Emma. "Making Big Promises Come True? Articulating and Realizing Value in Synthetic Biology." *BioSocieties* 8, no. 4 (2013): 432–48.

Frow, Emma, and Jane Calvert. "'Can Simple Biological Systems Be Built from Standardized Interchangeable Parts?': Negotiating Biology and Engineering in a Synthetic Biology Competition." *Engineering Studies* 5, no. 1 (2013): 42–58.

Galison, Peter. "Aufbau/Bauhaus: Logical Positivism and Architectural Modernism." *Critical Inquiry* 16, no. 4 (1990): 709–52.

———. *Einstein's Clocks, Poincaré's Maps: Empires of Time.* New York: W. W. Norton, 2004.

Galison, Peter, and Bruce Hevly. *Big Science: The Growth of Large-Scale Research.* Stanford, CA: Stanford University Press, 1992.

Galton, Francis. *Hereditary Genius: An Inquiry into Its Laws and Consequences.* New York: D. Appleton, 1870.

Gibbs, Nancy. "Creation Myths." *Time*, June 28, 2010.

Gibson, Daniel G., Gwynedd A. Benders, Cynthia Andrews-Pfannkoch, Evgeniya A. Denisova, Holly Baden-Tillson, Jayshree Zaveri, Timothy B. Stockwell, et al. "Complete Chemical Synthesis, Assembly, and Cloning of a *Mycoplasma genitalium* Genome." *Science* 319, no. 5867 (2008): 1215–20.

Gieryn, Thomas F. "Boundary-Work and the Demarcation of Science from Nonscience: Strains and Interests in Professional Ideologies of Scientists." *American Sociological Review* 48, no. 6 (1983): 781–95.

———. *Cultural Boundaries of Science: Credibility on the Line.* Chicago: University of Chicago Press, 1999.

Ginsberg, Daisy. "The Synthetic Kingdom." *Second Nature*, March 2010.

Ginsberg, Daisy, Jane Calvert, Pablo Schyfter, Alistair Elfick, and Drew Endy. *Synthetic Aesthetics: Investigating Synthetic Biology's Designs on Nature.* Cambridge, MA: MIT Press, 2014.

"The Go-Nowhere Generation." *New York Times*, March 10, 2012, sec. Opinion / Sunday Review.

Gordin, Michael. *The Pseudoscience Wars: Immanuel Velikovsky and the Birth of the Modern Fringe.* Chicago: University of Chicago Press, 2012.

Gould, Stephen J. "The Panda's Peculiar Thumb." *Natural History* 87, no. 9 (1978): 20–30.

Greenberg, W., and D. Kamin. "Property Rights and Payment to Patients for Cell

Lines Derived from Human Tissues: An Economic Analysis." *Social Science and Medicine* 36, no. 8 (1993): 1071–76.

Greenhalgh, Paul. "The History of Craft." In *The Culture of Craft*, edited by Peter Dormer, 20–51. Manchester, UK: Manchester University Press, 1997.

Gris, Juan. *Bulletin de la vie artistique* 6, no. 1 (1925): 15–17.

Gross, Charles G. "Claude Bernard and the Constancy of the Internal Environment." *Neuroscientist* 4, no. 5 (1998): 380–85.

Gutmann, Amy, and James W. Wagner. *New Directions: The Ethics of Synthetic Biology and Emerging Technologies*. Presidential Commission for the Study of Bioethical Issues, Washington, DC, 2010. http://bioethics.gov/sites/default/files /PCSBI-Synthetic-Biology-Report-12.16.10_0.pdf.

Haeckel, Ernst. *Generelle Morphologie der Organismen*. Berlin: G. Reimer, 1866.

Hagen, Joel B. "The Development of Experimental Methods in Plant Taxonomy, 1920–1950." *Taxon* 32, no. 3 (1983): 406–16.

———. "Ecologists and Taxonomists: Divergent Traditions in Twentieth-Century Plant Geography." *Journal of the History of Biology* 19, no. 2 (1986): 197–214.

———. "Experimentalists and Naturalists in Twentieth-Century Botany: Experimental Taxonomy, 1920–1950." *Journal of the History of Biology* 17, no. 2 (1984): 249–70.

———. "Naturalists, Molecular Biologists, and the Challenges of Molecular Evolution." *Journal of the History of Biology* 32, no. 2 (1999): 321–41.

———. "The Origins of Bioinformatics." *Nature Reviews: Genetics* 1, no. 3 (2000): 231–36.

———. "Retelling Experiments: H. B. D. Kettlewell's Studies of Industrial Melanism in Peppered Moths." *Biology and Philosophy* 14, no. 1 (1999): 39–54.

Hagstrom, Warren O. *The Scientific Community*. New York: Basic Books, 1965.

Hamilton, William. "ART. IX.-1. Artis Logicoe Rudimenta, with Illustrative Observations on Each Section." *Edinburgh Review, 1802–1929* 57, no. 115 (1833): 194–238.

Handley, Susannah. *Nylon, the Story of a Fashion Revolution: A Celebration of Design from Art Silk to Nylon and Thinking Fibres*. Baltimore: Johns Hopkins University Press, 1999.

Hansen, Birgitte Gorm. "Adapting in the Knowledge Economy: Lateral Strategies for Scientists and Those Who Study Them." PhD diss., School of Economics and Management, 2011.

Haraway, Donna J. *Modest_Witness@Second_Millennium.FemaleMan_Meets_ OncoMouse: Feminism and Technoscience*. New York: Routledge, 1997.

———. *Primate Visions: Gender, Race, and Nature in the World of Modern Science*. New York: Routledge, 1989.

———. *Simians, Cyborgs, and Women: The Reinvention of Nature*. New York: Routledge, 1991.

Hayden, Cori. "Gender, Genetics, and Generation: Reformulating Biology in Lesbian Kinship." *Cultural Anthropology* 10, no. 1 (1995): 41–63.

———. "Kinship Theory, Property, and the Politics of Inclusion: From Lesbian Families to Bioprospecting in a Few Short Steps." *Signs: Journal of Women in Culture and Society* 32, no. 2 (2007): 337–45.

———. *When Nature Goes Public: The Making and Unmaking of Bioprospecting in Mexico.* Princeton, NJ: Princeton University Press, 2003.

Hayles, N. Katherine. *My Mother Was a Computer: Digital Subjects and Literary Texts.* Chicago: University of Chicago Press, 2010.

Helmreich, Stefan. *Alien Ocean: An Anthropology of Marine Microbiology and the Limits of Life.* Berkeley: University of California Press, 2009.

———. "Flexible Infections: Computer Viruses, Human Bodies, Nation-States, Evolutionary Capitalism." *Science, Technology, and Human Values* 25, no. 4 (2000): 472–91.

———. "Nature/Culture/Seawater." *American Anthropologist* 113, no. 1 (2011): 132–44.

———. *Silicon Second Nature: Culturing Artificial Life in a Digital World.* Berkeley: University of California Press, 1998.

———. "Species of Biocapital." *Science as Culture* 17, no. 4 (2008): 463–78.

———. "The Spiritual in Artificial Life: Recombining Science and Religion in a Computational Culture Medium." *Science as Culture* 6, no. 3 (1997): 363–95.

———. "What Was Life? Answers from Three Limit Biologies." *Critical Inquiry* 37, no. 4 (2011): 671–96.

Helmreich, Stefan, and Sophia Roosth. "Life Forms: A Keyword Entry." *Representations* 112, no. 1 (2010): 27–53.

Hesse, Mary B. *Models and Analogies in Science.* Notre Dame, IN: University of Notre Dame Press, 1966.

Higgins, Dick. "Fluxus: Theory and Reception." Paper presented at "Fluxus: A Workshop Series, Alternative Traditions in the Contemporary Arts," University of Iowa, Iowa City, April 1985.

Highfield, Roger. "From Reading DNA to Writing It." *Telegraph*, June 28, 2007, sec. World News.

Hodgman, C. Eric, and Michael C. Jewett. "Cell-Free Synthetic Biology: Thinking Outside the Cell." *Metabolic Engineering* 14, no. 3 (2012): 261–69.

Hoefinger, Heidi. "'Professional Girlfriends.'" *Cultural Studies* 25, no. 2 (2011): 244–66.

Holmes, Frederic L. "Claude Bernard, the 'Milieu Intérieur,' and Regulatory Physiology." *History and Philosophy of the Life Sciences* 8, no. 1 (1986): 3–25.

———. "The Old Martyr of Science: The Frog in Experimental Physiology." *Journal of the History of Biology* 26, no. 2 (1993): 311–28.

Holmes, Frederic L., and William C. Summers. *Reconceiving the Gene: Seymour Benzer's Adventures in Phage Genetics.* New Haven, CT: Yale University Press, 2006.

Holt, Sharon Ann. "History Keeps Bethlehem Steel from Going off the Rails: Moving a Complex Community Process toward Success." *Public Historian* 28, no. 2 (2006): 31–44.

Hooke, Robert. "Discourse of Earthquakes." In *The Posthumous Works of Robert Hooke, M.D. S.R.S. Geom. Prof. Gresh., Etc., Containing His Cutlerian Lectures, and Other Discourses, Read at the Meetings of the Illustrious Royal Society*, 279–328. London, 1705.

Horgan, John. *The End of Science: Facing the Limits of Knowledge in the Twilight of the Scientific Age.* Reading, MA: Helix Books, 1996.

Hounshell, David A. *From the American System to Mass Production, 1800–1932: The Development of Manufacturing Technology in the United States.* Baltimore: Johns Hopkins University Press, 1985.

Hounshell, David A., and John Kenly Smith. *Science and Corporate Strategy: DuPont R&D, 1902–1980.* Cambridge: Cambridge University Press, 1988.

Hsu, Elisabeth. "Reflections on the 'Discovery' of the Antimalarial Qinghao." *British Journal of Clinical Pharmacology* 61, no. 6 (2006): 666–70.

Hughes, Sally Smith. *Genentech: The Beginnings of Biotech.* Chicago: University of Chicago Press, 2013.

———. "Making Dollars out of DNA: The First Major Patent in Biotechnology and the Commercialization of Molecular Biology, 1974–1980." *Isis* 92, no. 3 (2001): 541–75.

Hütter, Ralf, and Florian Schneider. Interview with Kraftwerk by Paul Alessandrini, November 1976. http://www.thing.de/delektro/www-eng/kw11-76.html.

Irigaray, Luce. *Speculum of the Other Woman.* Ithaca, NY: Cornell University Press, 1985.

"Is This Really the Worst Economic Recovery since the Depression?" *Economix Blog*, August 10, 2012. http://economix.blogs.nytimes.com/2012/08/10/is-this-really-the-worst-economic-recovery-since-the-depression/.

Jackson, Andrew. "Labour as Leisure—the Mirror Dinghy and DIY Sailors." *Journal of Design History* 19, no. 1 (2006): 57–67.

Jacob, François. *The Logic of Life: A History of Heredity.* New York: Vintage, 1976.

Jardine, Nicholas, James A. Secord, and Emma C. Spary. *Cultures of Natural History.* Cambridge: Cambridge University Press, 1996.

Jha, Alok. "From the Cells Up." *Guardian*, March 10, 2005, sec. Science.

Johns, Adrian. *The Nature of the Book.* Chicago: University of Chicago Press, 1998.

———. *Piracy: The Intellectual Property Wars from Gutenberg to Gates.* Chicago: University of Chicago Press, 2011.

Joyce, James. *A Portrait of the Artist as a Young Man.* New York: B. W. Huebsch, 1916.

Kahnweiler, Daniel-Henry. *The Rise of Cubism.* Translated by Henry Aronson. New York: Wittenborn, 1949.

Kanigel, Robert. *Faux Real: Genuine Leather and 200 Years of Inspired Fakes.* Philadelphia: University of Pennsylvania Press, 2010.

Kant, Immanuel. *Critique of Pure Reason.* Indianapolis: Hackett, 1996.

Kay, Lily. *The Molecular Vision of Life: Caltech, the Rockefeller Foundation, and the Rise of the New Biology*. New York: Oxford University Press, 1993.

———. *Who Wrote the Book of Life? A History of the Genetic Code*. Stanford, CA: Stanford University Press, 2000.

Keilin, David. "The Leeuwenhoek Lecture: The Problem of Anabiosis or Latent Life: History and Current Concept." *Proceedings of the Royal Society of London, Series B—Biological Sciences* 150, no. 939 (1959): 149–91.

Keller, Evelyn Fox. *Making Sense of Life: Explaining Biological Development with Models, Metaphors, and Machines*. Cambridge, MA: Harvard University Press, 2002.

———. *The Mirage of a Space between Nature and Nurture*. Durham, NC: Duke University Press, 2010.

———. "Organisms, Machines, and Thunderstorms: A History of Self-Organization, Part One." *Historical Studies in the Natural Sciences* 38, no. 1 (2008): 45–75.

———. "Organisms, Machines, and Thunderstorms: A History of Self-Organization, Part Two." *Historical Studies in the Natural Sciences* 39, no. 1 (2009): 1–31.

———. *Refiguring Life: Metaphors of Twentieth-Century Biology*. New York: Columbia University Press, 1995.

———. *Secrets of Life, Secrets of Death: Essays on Language, Gender, and Science*. New York: Routledge, 1992.

———. "What Does Synthetic Biology Have to Do with Biology?" *BioSocieties* 4, nos. 2–3 (2009): 291–302.

Kelty, Christopher M. "Free Software / Free Science." *First Monday* 6, no. 12 (2001). http://firstmonday.org/htbin/cgiwrap/bin/ojs/index.php/fm/article/view/902/811.

———. *Two Bits: The Cultural Significance of Free Software and the Internet*. Durham, NC: Duke University Press, 2008.

Kera, Denisa. "Innovation Regimes Based on Collaborative and Global Tinkering: Synthetic Biology and Nanotechnology in the Hackerspaces." *Technology in Society* 37 (2014): 28–37.

Kevles, Daniel J. "Ananda Chakrabarty Wins a Patent: Biotechnology, Law, and Society, 1972–1980." *Historical Studies in the Physical and Biological Sciences* 25, no. 1 (1994): 111–35.

———. *The Code of Codes: Scientific and Social Issues in the Human Genome Project*. Cambridge, MA: Harvard University Press, 1992.

———. "Of Mice and Money: The Story of the World's First Animal Patent." *Daedalus* 131, no. 2 (2002): 78–88.

———. "Patents, Protections, and Privileges: The Establishment of Intellectual Property in Animals and Plants." *Isis* 98, no. 2 (2007): 323–31.

———. "Protections, Privileges, and Patents: Intellectual Property in American Horticulture, the Late Nineteenth Century to 1930." *Proceedings of the American Philosophical Society* 152, no. 2 (2008): 207–13.

Klein, Ursula. *Experiments, Models, Paper Tools: Cultures of Organic Chemistry in the Nineteenth Century*. Stanford, CA: Stanford University Press, 2003.

Klein, Ursula, ed. *Tools and Modes of Representation in the Laboratory Sciences*. London: Kluwer, 2001.

Kloppenburg, Jack Ralph. *First the Seed: The Political Economy of Plant Biotechnology, 1492–2000*. Cambridge: Cambridge University Press, 1990.

Koerner, Lisbet. *Linnaeus: Nature and Nation*. Cambridge, MA: Harvard University Press, 2009.

Kohler, Robert E. *Landscapes and Labscapes: Exploring the Lab-Field Border in Biology*. Chicago: University of Chicago Press, 2002.

———. *Lords of the Fly: Drosophila Genetics and the Experimental Life*. Chicago: University of Chicago Press, 1994.

Krimsky, Sheldon. "Social Responsibility in an Age of Synthetic Biology." *Environment: Science and Policy for Sustainable Development* 24, no. 6 (1982): 2–5.

Labinger, Jay A., and Harry M. Collins. *The One Culture? A Conversation about Science*. Chicago: University of Chicago Press, 2001.

Lai, Larissa. *Salt Fish Girl: A Novel*. Toronto, ON: Thomas Allen, 2002.

Landecker, Hannah. "Between Beneficence and Chattel: The Human Biological in Law and Science." *Science in Context* 12, no. 1 (1999): 203–25.

———. *Culturing Life: How Cells Became Technologies*. Cambridge, MA: Harvard University Press, 2007.

———. "New Times for Biology: Nerve Cultures and the Advent of Cellular Life in Vitro." *Studies in History and Philosophy of Science Part C: Studies in History and Philosophy of Biological and Biomedical Sciences* 33, no. 4 (2002): 667–94.

Latour, Bruno. "A Cautious Prometheus? A Few Steps toward a Philosophy of Design (with Special Attention to Peter Sloterdijk)." Keynote lecture, Networks of Design Meeting, Falmouth, Cornwall, September 3, 2008. In *Proceedings of the 2008 Annual International Conference of the Design History Society*. http://www.bruno-latour.fr/sites/default/files/112-DESIGN-CORNWALL-GB.pdf.

Latour, Bruno, and Steve Woolgar. *Laboratory Life: The Construction of Scientific Facts*. Princeton, NJ: Princeton University Press, 1986.

Lazzarato, Maurizio. "From Capital-Labour to Capital-Life." *Ephemera: Theory and Politics in Organization* 4, no. 3 (2004): 187–208.

———. "Immaterial Labor." In *Radical Thought in Italy: A Potential Politics*, edited by Paolo Virno and Michael Hardt, 133–50. Minneapolis: University of Minnesota Press, 1996.

Leach, Robert. *Revolutionary Theatre*. New York: Routledge, 1994.

Lears, T. J. Jackson. *No Place of Grace: Antimodernism and the Transformation of American Culture, 1880–1920*. Chicago: University of Chicago Press, 1994.

Leduc, Stéphane. *La biologie synthétique*. Paris: A. Poinat, 1912.

————. *The Mechanism of Life*. London: William Heinemann, 1911.

Lee, Paula Young. "The Meaning of Molluscs: Leonce Reynaud and the Cuvier-Geoffroy Debate of 1830, Paris." *Journal of Architecture* 3, no. 3 (1998): 211–40.

Lenoir, Timothy. *Inscribing Science: Scientific Texts and the Materiality of Communication*. Stanford, CA: Stanford University Press, 1998.

————. *The Strategy of Life: Teleology and Mechanics in Nineteenth-Century German Biology*. Chicago: University of Chicago Press, 1989.

Leslie, Esther. *Synthetic Worlds: Nature, Art and the Chemical Industry*. London: Reaktion, 2005.

Levin, Thomas Y. " 'Tones from out of Nowhere': Rudolph Pfenninger and the Archaeology of Synthetic Sound." *Grey Room* 12 (2003): 32–79.

Levy, Steven. *Hackers: Heroes of the Computer Revolution*. New York: Basic Books, 1984.

Leyshon, Cressida. "This Week in Fiction: Thomas Pierce." *New Yorker Blogs*, December 17, 2012. http://www.newyorker.com/online/blogs/books/2012/12/this-week-in-fiction-thomas-pierce.html.

Liebig, Justus von. *Lettres sur la chimie*. Paris: Librairie Baillère, 1845.

Linné, Carl von. *Systema Naturae: Per Regna Tria Naturae, Secundum Classes, Ordines, Genera, Species, Cum Characteribus, Differentiis, Synonomis, Locis*. Beer, 1735.

Loeb, Jacques. *The Organism as a Whole: From a Physicochemical Viewpoint*. New York: Putnam's Sons, 1916.

Lopukhov, Fedor. *Writings on Ballet and Music*. Madison: University of Wisconsin Press, 2002.

Mackenzie, Adrian. "From Validating to Verifying: Public Appeals in Synthetic Biology." *Science as Culture* 22, no. 4 (2013): 476–96.

Mackenzie, Adrian, Claire Waterton, Rebecca Ellis, Emma Frow, Ruth McNally, Lawrence Busch, and Brian Wynne. "Classifying, Constructing, and Identifying Life Standards as Transformations of 'the Biological.' " *Science, Technology, and Human Values* 38, no. 5 (2013): 701–22.

MacPhail, Theresa M. "The Viral Gene: An Undead Metaphor Recoding Life." *Science as Culture* 13, no. 3 (2004): 325–45.

Malinowski, Bronisław. *Argonauts of the Western Pacific: An Account of Native Enterprise and Adventure in the Archipelagoes of Melanesian New Guinea*. New York: G. Routledge and Sons, 1922.

Marcus, George E., and Michael M. J. Fischer. *Anthropology as Cultural Critique: An Experimental Moment in the Human Sciences*. Chicago: University of Chicago Press, 1986.

Mardis, Elaine R. "A Decade's Perspective on DNA Sequencing Technology." *Nature* 470, no. 7333 (2011): 198–203.

Marinetti, Filippo Tommaso, Emilo Settimelli, and Bruno Corra. "The Synthetic Futurist Theatre." In *Let's Murder the Moonshine: Selected Writings*, edited by

R. W. Flint, translated by R. W. Flint and Arthur A. Coppotelli, 135. Los Angeles: Sun and Moon Classics, 1991.

Martin, Collin H., David R. Nielsen, Kevin V. Solomon, and Kristala L. Jones Prather. "Synthetic Metabolism: Engineering Biology at the Protein and Pathway Scales." *Chemistry and Biology* 16, no. 3 (2009): 277–86.

Marx, Karl. "The 18th Brumaire of Louis Bonaparte." In *Selected Works*, by Karl Marx and Frederick Engels, vol. 1. Moscow: Progress Publishers, 1969.

———. *Grundrisse: Foundations of the Critique of Political Economy*. New York: Random House, 1973.

———. *Karl Marx: Selected Writings*. Oxford: Oxford University Press, 2000.

Masami, Nomura. "Myths of the Toyota System." *AMPO: Japan-Asia Quarterly* 25, no. 1 (1994): 18–25.

Maurer, Bill. *Mutual Life, Limited: Islamic Banking, Alternative Currencies, Lateral Reason*. Princeton, NJ: Princeton University Press, 2005.

Mayr, Ernst. *The Evolutionary Synthesis: Perspectives on the Unification of Biology, with a New Preface*. Cambridge, MA: Harvard University Press, 1998.

———. *Systematics and the Origin of Species, from the Viewpoint of a Zoologist*. Cambridge, MA: Harvard University Press, 1942.

McCray, Patrick. *Keep Watching the Skies! The Story of Operation Moonwatch and the Dawn of the Space Age*. Princeton, NJ: Princeton University Press, 2008.

"Meanings of 'Life.'" *Nature* 447, no. 7148 (2007): 1031–32.

Meek, James, and Michael Ellison. "On the Path of Biology's Holy Grail: A Supercomputer Called Blue Gene Aims to Win the Race to Decode 'Book of Life.'" *Guardian*, June 5, 2000.

Metcalf, Bruce, ed. "Replacing the Myth of Modernism." In *Neocraft: Modernity and Crafts*, 4–32. Halifax: Press of the Nova Scotia College of Art and Design, 2007.

Metlay, Grischa. "Reconsidering Renormalization: Stability and Change in 20th-Century Views on University Patents." *Social Studies of Science* 36, no. 4 (2006): 565–97.

Meyer, Morgan. "Domesticating and Democratizing Science: A Geography of Do-It-Yourself Biology." *Journal of Material Culture* 18, no. 2 (2013): 117–34.

Milburn, Colin. "Just for Fun: The Playful Image of Nanotechnology." *NanoEthics* 5, no. 2 (2011): 223–32.

Milman, Estera. "Futurism as a Submerged Paradigm for Artistic Activism and Practical Anarchism." *South Central Review* 13, no. 2/3 (1996): 157–79.

Milton, John. *Areopagitica, and Other Prose Works*. Whitefish, MT: Kessinger, 2004.

Mirowski, Philip. *Science-Mart: Privatizing American Science*. Cambridge, MA: Harvard University Press, 2011.

Mitchell, Robert, and Philip Thurtle, eds. *Data Made Flesh: Embodying Information*. New York: Routledge, 2004.

Mitchell, Timothy. *Rule of Experts: Egypt, Techno-politics, Modernity*. Berkeley: University of California Press, 2002.

Mody, Cyrus C. M. "Crafting the Tools of Knowledge: The Invention, Spread, and Commercialization of Probe Microscopy, 1960–2000." PhD diss., Cornell University, 2004.

Monod, Jacques, and François Jacob. "General Conclusions: Teleonomic Mechanisms in Cellular Metabolism, Growth, and Differentiation." *Cold Spring Harbor Symposia on Quantitative Biology (Cellular Regulatory Mechanisms)* 26 (1961): 389–401.

Moore, Fred. *Homebrew Computer Club Newsletter*, March 15, 1975. http://www.digibarn.com/collections/newsletters/homebrew/V1_01/index.html.

Morgan, Mary S., and Margaret Morrison. *Models as Mediators: Perspectives on Natural and Social Science*. Cambridge: Cambridge University Press, 1999.

Moses, Alan. "Intelligent Design: Playing with the Building Blocks of Biology." *Berkeley Science Review* 5, no. 1 (2005): 34–40.

Mukunda, Gautam, Kenneth A. Oye, and Scott C. Mohr. "What Rough Beast?" *Politics and the Life Sciences* 28, no. 2 (2009): 2–26.

Myers, Natasha. "Modeling Proteins, Making Scientists: An Ethnography of Pedagogy and Visual Cultures in Contemporary Structural Biology." PhD diss., Massachusetts Institute of Technology, 2007.

———. "Molecular Embodiments and the Body-Work of Modeling in Protein Crystallography." *Social Studies of Science* 38, no. 2 (2008): 163–99.

———. *Rendering Life Molecular: Models, Modelers, and Excitable Matter*. Durham, NC: Duke University Press, 2015.

Ndiaye, Pap. *Nylon and Bombs: DuPont and the March of Modern America*. Baltimore: Johns Hopkins University Press, 2007.

Negri, Antonio. "Constituent Republic." *Common Sense* 16 (1994): 89–96.

Nelkin, Dorothy. "Genetics, God, and Sacred DNA." *Society* 33, no. 4 (1996): 22–25.

"The 1918 Flu Virus Is Resurrected." *Nature* 437, no. 7060 (2005): 794–95.

Noble, David F. *Forces of Production: A Social History of Industrial Automation*. New York: Oxford University Press, 1986.

———. *Progress without People: New Technology, Unemployment, and the Message of Resistance*. Toronto: Between the Lines, 1995.

Normandin, Sebastian. "Claude Bernard and *An Introduction to the Study of Experimental Medicine*: 'Physical Vitalism,' Dialectic, and Epistemology." *Journal of the History of Medicine and Allied Sciences* 62, no. 4 (2007): 495–528.

Novas, Carlos, and Nikolas Rose. "Genetic Risk and the Birth of the Somatic Individual." *Economy and Society* 29, no. 4 (2000): 485–513.

Numbers, Ronald L. *Darwinism Comes to America*. Cambridge, MA: Harvard University Press, 1998.

Nutch, Frank. "Gadgets, Gizmos, and Instruments: Science for the Tinkering." *Science, Technology, and Human Values* 21, no. 2 (1996): 214–28.

Nye, David E. *America's Assembly Line*. Cambridge, MA: MIT Press, 2013.

Nyhart, Lynn K. *Biology Takes Form: Animal Morphology and the German Universities, 1800–1900*. Chicago: University of Chicago Press, 1995.

————. *Modern Nature: The Rise of the Biological Perspective in Germany*. Chicago: University of Chicago Press, 2009.

O'Connell, Mark. "Has James Joyce Been Set Free?" *New Yorker Blogs*, January 11, 2012. http://www.newyorker.com/online/blogs/books/2012/01/james-joyce-public-domain.html.

Oldham, Paul, Stephen Hall, and Geoff Burton. "Synthetic Biology: Mapping the Scientific Landscape." *PLoS ONE* 7, no. 4 (2012): e34368.

Oreskes, Naomi, and Erik M. Conway. *Merchants of Doubt: How a Handful of Scientists Obscured the Truth on Issues from Tobacco Smoke to Global Warming*. New York: Bloomsbury Press, 2010.

Ospovat, Dov. "God and Natural Selection: The Darwinian Idea of Design." *Journal of the History of Biology* 13, no. 2 (1980): 169–94.

Oyama, Susan. *The Ontogeny of Information: Developmental Systems and Evolution*. Durham, NC: Duke University Press, 2000.

Paley, William, and James Paxton. *Natural Theology; or, Evidences of the Existence and Attributes of the Deity, Collected from the Appearances of Nature*. Boston: Gould and Lincoln, 1854.

Parry, Bronwyn. "Technologies of Immortality: The Brain on Ice." *Studies in History and Philosophy of Science Part C: Studies in History and Philosophy of Biological and Biomedical Sciences* 35, no. 2 (2004): 391–413.

Patterson, Meredith. "A Biopunk Manifesto." Presented at the UCLA Center for Society and Genetics symposium "Outlaw Biology? Public Participation in the Age of Big Bio," January 29, 2010. http://maradydd.livejournal.com/496085.html.

Pauly, Philip J. *Controlling Life: Jacques Loeb and the Engineering Ideal in Biology*. Oxford: Oxford University Press, 1987.

Paxson, Heather. *The Life of Cheese: Crafting Food and Value in America*. Berkeley: University of California Press, 2013.

Pehnt, Wolfgang. "Gropius the Romantic." *Art Bulletin* 53, no. 3 (1971): 379–92.

Pennisi, Elizabeth. "Synthetic Biology Remakes Small Genomes." *Science* 310, no. 5749 (2005): 769–70.

Perlman, Robert L. "The Concept of the Organism in Physiology." *Theory in Biosciences* 119, nos. 3–4 (2000): 174–86.

Petryna, Adriana. *When Experiments Travel: Clinical Trials and the Global Search for Human Subjects*. Princeton, NJ: Princeton University Press, 2009.

Pickstone, John V. "Sketching Together the Modern Histories of Science, Technology, and Medicine." *Isis* 102, no. 1 (2011): 123–33.

Pierce, Thomas. "Shirley Temple Three." *New Yorker*, December 24, 2012.

Pinch, Trevor J., and Frank Trocco. *Analog Days: The Invention and Impact of the Moog Synthesizer*. Cambridge, MA: Harvard University Press, 2004.

Polanyi, Michael. *Personal Knowledge: Towards a Post-critical Philosophy*. New York: Routledge, 1962.

Popper, Karl. *The Logic of Scientific Discovery*. New York: Basic Books, 1959.

Pottage, Alain. "Too Much Ownership: Bio-prospecting in the Age of Synthetic Biology." *BioSocieties* 1, no. 2 (2006): 137–58.

Powers, Richard. *Orfeo*. New York: W. W. Norton, 2014.

Preston, Richard. *Panic in Level 4: Cannibals, Killer Viruses, and Other Journeys to the Edge of Science*. New York: Random House, 2008.

Proctor, Robert. *Cancer Wars: How Politics Shapes What We Know and Don't Know about Cancer*. New York: Basic Books, 1995.

———. *Golden Holocaust: Origins of the Cigarette Catastrophe and the Case for Abolition*. Berkeley: University of California Press, 2012.

Proctor, Robert, and Londa L. Schiebinger. *Agnotology: The Making and Unmaking of Ignorance*. Stanford, CA: Stanford University Press, 2008.

Prusinkiewicz, Przemyslaw, and Aristid Lindenmayer. *The Algorithmic Beauty of Plants*. Berlin: Springer, 1990.

Purnick, Priscilla E. M., and Ron Weiss. "The Second Wave of Synthetic Biology: From Modules to Systems." *Nature Reviews: Molecular Cell Biology* 10, no. 6 (2009): 410–22.

Putnam, F. W., D. Miller, L. Palm, and E. A. Evans. "Biochemical Studies of Virus Reproduction." *Journal of Biological Chemistry* 199, no. 1 (1952): 177–91.

Rabinow, Paul. "Artificiality and Enlightenment: From Sociobiology to Biosociality." In *Zone 6: Incorporations*, edited by Jonathan Crary and Sanford Kwinter, 190–201. New York: Zone Books, 1992.

———. *Making PCR: A Story of Biotechnology*. Chicago: University of Chicago Press, 1996.

Rader, Karen A. *Making Mice: Standardizing Animals for American Biomedical Research, 1900–1955*. Princeton, NJ: Princeton University Press, 2004.

Radin, Joanna. "Latent Life: Concepts and Practices of Human Tissue Preservation in the International Biological Program." *Social Studies of Science* 43, no. 4 (2013): 484–508.

Rai, Arti, and James Boyle. "Synthetic Biology: Caught between Property Rights, the Public Domain, and the Commons." *PLoS Biology* 5, no. 3 (2007): e58.

Rajagopal, Deepak, Steve Sexton, Gal Hochman, and David Zilberman. "Recent Developments in Renewable Technologies: R&D Investment in Advanced Biofuels." *Annual Review of Resource Economics* 1, no. 1 (2009): 621–44.

Rapp, Rayna. *Testing Women, Testing the Fetus: The Social Impact of Amniocentesis in America*. New York: Routledge, 1999.

Rapp, Rayna, Deborah Heath, and Karen-Sue Taussig. "Genealogical Dis-ease: Where Hereditary Abnormality, Biomedical Explanation, and Family Responsibility Meet." In *Relative Values: Reconfiguring Kinship Studies*, edited by Sarah Franklin and Susan McKinnon, 384–409. Durham, NC: Duke University Press, 2001.

Rasmussen, Nicolas. *Picture Control: The Electron Microscope and the Transformation of Biology in America, 1940–1960*. Stanford, CA: Stanford University Press, 1997.

Ray, Thomas S. "An Evolutionary Approach to Synthetic Biology: Zen and the Art of Creating Life." *Artificial Life* 1, no. 1/2 (1993): 179–209.

Reardon, Jenny. *Race to the Finish: Identity and Governance in an Age of Genomics.* Princeton, NJ: Princeton University Press, 2004.

Reynolds, Andrew. "The Cell's Journey: From Metaphorical to Literal Factory." *Endeavour* 31, no. 2 (2007): 65–70.

Rheinberger, Hans-Jörg. *Toward a History of Epistemic Things: Synthesizing Proteins in the Test Tube.* Stanford, CA: Stanford University Press, 1997.

Richards, Robert J. *The Meaning of Evolution: The Morphological Construction and Ideological Reconstruction of Darwin's Theory.* Chicago: University of Chicago Press, 1993.

———. *The Romantic Conception of Life: Science and Philosophy in the Age of Goethe.* Chicago: University of Chicago Press, 2002.

———. *The Tragic Sense of Life: Ernst Haeckel and the Struggle over Evolutionary Thought.* Chicago: University of Chicago Press, 2008.

Riskin, Jessica. "The Divine Optician." *American Historical Review* 116, no. 2 (2011): 352–70.

———. *Genesis Redux: Essays in the History and Philosophy of Artificial Life.* Chicago: University of Chicago Press, 2007.

Ritvo, Harriet. "Animal Planet." *Environmental History* 9, no. 2 (2004): 204–20.

———. "Beasts in the Jungle (or Wherever)." *Daedalus* 137, no. 2 (2008): 22–30.

———. *The Platypus and the Mermaid, and Other Figments of the Classifying Imagination.* Cambridge, MA: Harvard University Press, 1998.

———. "Race, Breed, and Myths of Origin: Chillingham Cattle as Ancient Britons." *Representations*, no. 39 (1992): 1–22.

Ro, Dae-Kyun, Eric M. Paradise, Mario Ouellet, Karl J. Fisher, Karyn L. Newman, John M. Ndungu, Kimberly A. Ho, et al. "Production of the Antimalarial Drug Precursor Artemisinic Acid in Engineered Yeast." *Nature* 440, no. 7086 (2006): 940–43.

Robbins, Daniel. "Abbreviated Historiography of Cubism." *Art Journal* 47, no. 4 (1988): 277–83.

Roberts, Leslie. "Tough Times Ahead for the Genome Project." *Science* 248, no. 4963 (1990): 1600–1601.

Roblin, R. "Synthetic Biology." *Nature* 282 (1979): 171–72.

Roosth, Sophia. "The Godfather, Part II." *Science* 342, no. 6156 (2013): 312–13.

———. "Life, Not Itself: Inanimacy and the Limits of Biology." *Grey Room* 57, no. 3 (2014): 56–81.

Rose, Mark. *Authors and Owners: The Invention of Copyright.* Cambridge, MA: Harvard University Press, 1995.

———. "Mothers and Authors: *Johnson v. Calvert* and the New Children of Our Imaginations." *Critical Inquiry* 22, no. 4 (1996): 613–33.

Ross, Andrew. *Science Wars.* Durham, NC: Duke University Press, 1996.

Rubin, Gayle. "The Traffic in Women: Notes on the 'Political Economy' of Sex."

In *Toward an Anthropology of Women*, edited by Rayna R. Reiter, 157–210. New York: Monthly Review Press, 1975.

Rupke, Nicolaas A. *Richard Owen: Biology without Darwin*. Chicago: University of Chicago Press, 2009.

Sagmeister, Stefan, and Drew Endy. "On Design." *In Science Is Culture: Conversations at the New Intersection of Science and Society*, edited by Adam Bly, 61–74. New York: HarperCollins, 2010.

Salingaros, Nikos Angelos, and Christopher Alexander. *Anti-architecture and Deconstruction*. Solingen: Umbau-Verlag, 2008.

Salter, Chris. *Entangled: Technology and the Transformation of Performance*. Cambridge, MA: MIT Press, 2010.

Schacht, Wendy. *The Bayh-Dole Act: Selected Issues in Patent Policy and the Commercialization of Technology*. Congressional Research Service Report for Congress, December 3, 2012.

Schaffer, Simon. "Babbage's Intelligence: Calculating Engines and the Factory System." *Critical Inquiry* 21, no. 1 (1994): 203–27.

———. "Glass Works: Newton's Prisms and the Uses of Experiment." In *The Uses of Experiment*, edited by David Gooding, Trevor Pinch, and Simon Schaffer, 67–104. Cambridge: Cambridge University Press, 1989.

Schnapp, Jeffrey T. "The Fabric of Modern Times." *Critical Inquiry* 24, no. 1 (1997): 191–245.

Schneider, David M. *American Kinship: A Cultural Account*. 2nd ed. Chicago: University of Chicago Press, 1980. First published 1968.

Schneider, Jane. "In and out of Polyester: Desire, Disdain and Global Fibre Competitions." *Anthropology Today* 10, no. 4 (1994): 2–10.

Schrödinger, Erwin. *What Is Life? The Physical Aspect of the Living Cell*. New York: Macmillan, 1944.

Secord, Anne. "Science in the Pub: Artisan Botanists in Early Nineteenth-Century Lancashire." *History of Science: An Annual Review of Literature, Research and Teaching* 32, no. 97 (1994): 269–315.

Sedgwick, Eve Kosofsky. "Paranoid Reading and Reparative Reading; or, You're So Paranoid, You Probably Think This Introduction Is about You." In *Novel Gazing: Queer Readings in Fiction*, edited by Eve Kosofsky Sedgwick, 1–39. Durham, NC: Duke University Press, 1997.

Sennett, Richard. *The Craftsman*. New Haven, CT: Yale University Press, 2008.

Service, Robert F. "Rethinking Mother Nature's Choices." *Science* 315, no. 5813 (2007): 793.

Shapin, Steven. "The House of Experiment in Seventeenth-Century England." *Isis* 79, no. 3 (1988): 373–404.

———. *The Scientific Life: A Moral History of a Late Modern Vocation*. Chicago: University of Chicago Press, 2009.

"Shell's Brash Biofuels Partner." *Forbes, April 22, 2009*. http://www.forbes.com/2009/04/22/codexis-shell-biofuel-technology-breakthroughs-codexis.html.

Sheridan, Cormac. "Making Green." *Nature Biotechnology* 27, no. 12 (2009): 1074–76.

Shimokawa, Koichi. *The Japanese Automobile Industry: A Business History.* London: Athlone, 1994.

Siegel, Jonas. "Interview: Drew Endy." *Bulletin of the Atomic Scientists* 63, no. 3 (2007): 28–33.

Silverman, Chloe. *Understanding Autism: Parents, Doctors, and the History of a Disorder.* Princeton, NJ: Princeton University Press, 2012.

Simondon, Gilbert. "The Genesis of the Individual." In *Zone 6: Incorporations*, edited by Jonathan Crary and Sanford Kwinter, 296–319. New York: Zone Books, 1992.

Sinsheimer, Robert L. "Recombinant DNA—on Our Own." *Bioscience* 26, no. 10 (1976): 599.

Sismondo, Sergio. "Models, Simulations, and Their Objects." *Science in Context* 12, no. 2 (1999): 247–60.

Smith, Pamela H. *The Body of the Artisan: Art and Experience in the Scientific Revolution.* Chicago: University of Chicago Press, 2004.

Smocovitis, Vassiliki Betty. *Unifying Biology.* Princeton, NJ: Princeton University Press, 1996.

Snow, C. P. *The Two Cultures.* Cambridge: Cambridge University Press, 1993.

Sober, Elliott. "The Design Argument." In *The Blackwell Guide to the Philosophy of Religion*, edited by William E. Mann, 117–47. London: Blackwell, 2008.

Solecki, Mary, Anna Scodel, and Bob Epstein. "Advanced Biofuel Market Report 2013: Capacity through 2016." Environmental Entrepreneurs, San Francisco, CA, 2013. http://www.eesi.org/files/E2AdvancedBiofuelMarketReport2013.pdf.

Spary, Emma C. *Utopia's Garden: French Natural History from Old Regime to Revolution.* Chicago: University of Chicago Press, 2010.

Spivak, Gayatri. "Subaltern Studies: Deconstructing Historiography." In *Selected Subaltern Studies*, 3–32. New Delhi: Oxford University Press, 1988.

Sprinzak, D., and M. B. Elowitz. "Reconstruction of Genetic Circuits." *Nature* 438, no. 7067 (2005): 443–48.

Stacey, Jackie. *The Cinematic Life of the Gene.* Durham, NC: Duke University Press, 2010.

State of Alabama. "State Testing Sugar-Cane Production for Biofuel Industry." Press release. October 21, 2008. http://www.media.alabama.gov/pr/pr.aspx?id=999.

Stevens, Hallam. *Life out of Sequence: A Data-Driven History of Bioinformatics.* Chicago: University of Chicago Press, 2013.

Stevens, Peter F. "Species: Historical Perspectives." In *Keywords in Evolutionary Biology*, edited by Evelyn Fox Keller and Elisabeth Anne Lloyd, 302–11. Cambridge, MA: Harvard University Press, 1992.

Stewart, Matthew. "The Management Myth." *Atlantic*, June 2006.

Stokstad, Erik. "Conservation Biology: Ivory Ban Upheld." *Science* 328, no. 5974 (2010): 26.

Strasser, Bruno J. "Collecting Nature: Practices, Styles, and Narratives." *Osiris* 27, no. 1 (2012): 303–40.

———. "The Experimenter's Museum: GenBank, Natural History, and the Moral Economies of Biomedicine." *Isis* 102, no. 1 (2011): 60–96.

———. "Genbank: Natural History in the 21st Century?" *Science* 322, no. 5901 (2008): 537–38.

———. "Laboratories, Museums, and the Comparative Perspective: Alan A. Boyden's Quest for Objectivity in Serological Taxonomy, 1924–1962." *Historical Studies in the Natural Sciences* 40, no. 2 (2010): 149–82.

Strathern, Marilyn. *Kinship, Law, and the Unexpected: Relatives Are Always a Surprise*. Cambridge: Cambridge University Press, 2005.

———. "No Nature, No Culture: The Hagen Case." In *Nature, Culture and Gender*, edited by Carol MacCormack and Marilyn Strathern, 174–222. Cambridge: Cambridge University Press, 1980.

———. *Reproducing the Future: Anthropology, Kinship, and the New Reproductive Technologies*. New York: Routledge, 1992.

———. Review of *Families We Choose: Lesbians, Gays, Kinship*, by Kath Weston. *Man* 28, no. 1 (1993): 195–96.

Sunder Rajan, Kaushik. *Biocapital: The Constitution of Postgenomic Life*. Durham, NC: Duke University Press, 2006.

———. *Lively Capital: Biotechnologies, Ethics, and Governance in Global Markets*. Durham, NC: Duke University Press, 2012.

———. "Pharmaceutical Crises and Questions of Value: Terrains and Logics of Global Therapeutic Politics." *South Atlantic Quarterly* 111, no. 2 (2012): 321–46.

"Synthetic Biology, a New Paradigm for Biological Discovery." Beachhead Consulting, February 2006. http://www.researchandmarkets.com/reports/314528/synthetic _biology_a_new_paradigm_for_biological.

Taubenberger, Jeffery K., Ann H. Reid, Raina M. Lourens, Ruixue Wang, Guozhong Jin, and Thomas G. Fanning. "Characterization of the 1918 Influenza Virus Polymerase Genes." *Nature* 437, no. 7060 (2005): 889–93.

Taussig, Karen-Sue, Rayna Rapp, and Deborah Heath. "Flexible Eugenics: Discourses of Perfectibility in Late Twentieth-Century America." In *Genetic Nature/Culture: Anthropology and Science beyond the Two-Culture Divide*, edited by Allan Goodman, Deborah Heath, and M. Susan Lindee, 58–76. Berkeley: University of California Press, 2003.

Tawfik, Dan S. "Messy Biology and the Origins of Evolutionary Innovations." *Nature Chemical Biology* 6, no. 10 (2010): 692–96.

Taylor, Frederick Winslow. *The Principles of Scientific Management*. New York: Cosimo, 2006. First published 1911.

Taylor, Timothy Dean. *Strange Sounds: Music, Technology, and Culture*. New York: Routledge, 2001.

Thacker, Eugene. *Biomedia*. Minneapolis: University of Minnesota Press, 2004.

Thompson, Charis. "The Biotech Mode of Reproduction." Paper presented at "Animation and Cessation: Anthropological Perspectives on Changing Definitions of Life and Death in the Context of Biomedicine," School of American Research Advanced seminar, Santa Fe, NM, 2000.

———. *Making Parents: The Ontological Choreography of Reproductive Technologies.* Cambridge, MA: MIT Press, 2005.

Todes, Daniel P. *Pavlov's Physiology Factory: Experiment, Interpretation, Laboratory Enterprise.* Baltimore: Johns Hopkins University Press, 2002.

Torda, Thomas Joseph. "Tairov's 'Princess Brambilla': A Fantastic, Phantasmagoric 'Capriccio' at the Moscow Kamerny Theatre." *Theatre Journal* 32, no. 4 (1980): 488–98.

"Trends in Synthetic Biology Research Funding in the United States and Europe." Woodrow Wilson International Center for Scholars, June 2010. http://www.synbioproject.org/publications/researchfunding/.

Tritch, Teresa. "Still Crawling out of a Very Deep Hole." *New York Times*, April 7, 2012, sec. Opinion / Sunday Review.

Turner, Fred. *From Counterculture to Cyberculture: Stewart Brand, the Whole Earth Network, and the Rise of Digital Utopianism.* Chicago: University of Chicago Press, 2006.

Tyson, Neil deGrasse. "What's the Next Big Thing? Profile: Jay Keasling." *NOVA scienceNOW.* PBS, February 23, 2011.

Van den Belt, Henk, and Arie Rip. "The Nelson-Winter-Dosi Model and Synthetic Dye Chemistry." In *The Social Construction of Technological Systems: New Directions in the Sociology and History of Technology*, edited by Wiebe E. Bijker, Thomas Parke Hughes, and Trevor Pinch, 129–54. Cambridge, MA: MIT Press, 2012.

Van Doren, Davy, Stefan Koenigstein, and Thomas Reiss. "The Development of Synthetic Biology: A Patent Analysis." *Systems and Synthetic Biology* 7, no. 4 (2013): 209–20.

Venter, J. Craig. *Life at the Speed of Light: From the Double Helix to the Dawn of Digital Life.* New York: Viking, 2013.

———. "Multiple Personal Genomes Await." *Nature* 464, no. 7289 (2010): 676–77.

Volkin, Elliot, L. Astrachan, and Joan L. Countryman. "Metabolism of RNA Phosphorus in *Escherichia coli* Infected with Bacteriophage T7." *Virology* 6, no. 2 (1958): 545–55.

Vora, Kalindi. *Life Support: Biocapital and the New History of Outsourced Labor.* Minneapolis: University of Minnesota Press, 2015.

Wade, Nicholas. "Researchers Say They Created a 'Synthetic Cell.'" *New York Times*, May 20, 2010, sec. Science.

Wakayama, Sayaka, Hiroshi Ohta, Takafusa Hikichi, Eiji Mizutani, Takamasa Iwaki, Osami Kanagawa, and Teruhiko Wakayama. "Production of Healthy Cloned Mice from Bodies Frozen at −20°C for 16 Years." *Proceedings of the National Academy of Sciences* 105 (2008): 17318–322.

Waldby, Catherine. *The Visible Human Project: Informatic Bodies and Posthuman Medicine*. New York: Routledge, 2000.

Waldby, Catherine, and Robert Mitchell. *Tissue Economies: Blood, Organs, and Cell Lines in Late Capitalism*. Durham, NC: Duke University Press, 2006.

Walker, John F., and Daniel Stiles. "Consequences of Legal Ivory Trade." *Science* 328, no. 5986 (2010): 1633–34; author reply, 1634–35.

Wang, Harris H., Farren J. Isaacs, Peter A. Carr, Zachary Z. Sun, George Xu, Craig R. Forest, and George M. Church. "Programming Cells by Multiplex Genome Engineering and Accelerated Evolution." *Nature* 460, no. 7257 (2009): 894–98.

Warner, Michael. *The Trouble with Normal: Sex, Politics, and the Ethics of Queer Life*. New York: Free Press, 1999.

Wasser, Samuel, Joyce Poole, Phyllis Lee, Keith Lindsay, Andrew Dobson, John Hart, Iain Douglas-Hamilton, et al. "Conservation: Elephants, Ivory, and Trade." *Science* 327, no. 5971 (2010): 1331–32.

Wertheim, Margaret. *Physics on the Fringe: Smoke Rings, Circlons, and Alternative Theories of Everything*. New York: Walker, 2013.

Weston, Kath. *Families We Choose: Lesbians, Gays, Kinship*. New York: Columbia University Press, 1997.

———. "Forever Is a Long Time: Romancing the Real in Gay Kinship Ideologies." In *Naturalizing Power: Essays in Feminist Cultural Analysis*, edited by Sylvia Junko Yanagisako and Carol Delaney, 87–112. New York: Routledge, 1995.

"What's in a Name?" *Nature Biotechnology* 27, no. 12 (2009): 1071–73.

Whittaker, R. H. "New Concepts of Kingdoms of Organisms." *Science* 163, no. 3863 (1969): 150–60.

Witze, Alexandra. "Light in the Dark: Factory of Life; Synthetic Biologists Reinvent Nature with Parts, Circuits." *Science News* 183, no. 1 (2013): 24.

Woese, C. R., O. Kandler, and M. L. Wheelis. "Towards a Natural System of Organisms: Proposal for the Domains Archaea, Bacteria, and Eucarya." *Proceedings of the National Academy of Sciences* 87, no. 12 (1990): 4576–79.

Wong, Winnie Won Yin. *Van Gogh on Demand: China and the Readymade*. Chicago: University of Chicago Press, 2014.

World Health Organization. *World Malaria Report. Geneva, Switzerland: World Health Organization*, 2006.

Wright, Susan. "Recombinant DNA Technology and Its Social Transformation, 1972–1982." *Osiris* 2 (1986): 303–60.

Yeh, Brian J., and Wendell A. Lim. "Synthetic Biology: Lessons from the History of Synthetic Organic Chemistry." *Nature Chemical Biology* 3, no. 9 (2007): 521–25.

Zimmer, Carl. "Bringing Them Back to Life." *National Geographic,* November 2013.

Zimov, Sergey A. "Pleistocene Park: Return of the Mammoth's Ecosystem." *Science* 308, no. 5723 (2005): 796–98.

Index